**Stripe Press**

Ideas for progress
San Francisco, California
press.stripe.com

# WHERE IS MY FLYING CAR?

J. STORRS HALL

Also by J. Storrs Hall

*Beyond AI: Creating the Conscience of the Machine* (2007)

*Nanofuture: What's Next for Nanotechnology* (2005)

Published in the United States of America
by Stripe Press / Stripe Matter Inc.

Stripe Press
Ideas for progress
San Francisco, California
press.stripe.com

Printed by Hemlock in Canada
ISBN: 978-1-953953-18-6
Library of Congress Control Number: 2021945676
First Edition

# Table of Contents

# WHERE IS MY FLYING CAR?

J. STORRS HALL

# I

# Profiles
# of the
# Past

# 1

# The World
# of Tomorrow

*I read "Tom Swift and His Airship"; "Tom Swift and His
Electric Runabout"; "Tom Swift and His Submarine Boat"
and I said that's for me.*

—C. L. "Kelly" Johnson, *Kelly*

Fifty years ago, it was clear that flying cars were in our future. Or at least it was clear to any 10-year-old boy who read the stories of Tom Swift Jr. His "triphibian atomicar" was held in the air by "repellatron rays" and powered by an "atomic capsule," which could "change atomic energy directly into electricity."[1]

The original Tom Swift Sr. was introduced in 1910, and had adventures with gadgets like an "electric runabout" and a "photo telephone." He was modeled on Thomas Edison, who in real life was a popular hero to an extent little realized today.[2] Edison invented not just the electric light and the phonograph but also the industrial research laboratory; he produced a huge swath of the technologies that distinguished the 20th century from the 19th; and he had towns (e.g., Edison, New Jersey) named after him by a grateful public.

A hundred years ago, people knew that the future was going to be wonderful. And they were right. They lived through decades of spectacular accomplishments—cars, radios, airplanes, electricity illuminating homes and cities—and unparalleled economic growth. Half a century ago, around the time that Tom Swift Jr. took over the series, people had reason to believe the same.

Edison had inspired Tom Swift, who in turn inspired Kelly Johnson. Johnson would become the first team leader of Lockheed Skunk Works and one of the 20th century's most celebrated aeronautical engineers. Johnson was famous for high-end military aircraft, such as the F-104 Starfighter and SR-71 Blackbird. But he also introduced the first airliner with retractable landing gear and the first pressurized airliner to go into widespread use. By 1960 he and Lockheed had introduced the first dedicated business jet, the JetStar. The flying car didn't seem that far off.

The vision of a bright future world enabled by technology, which formed the background of the 1960s Space Age zeitgeist, had been growing in collective imaginations since the turn of the century. It accelerated, rather than slowing, during the Great Depression; perhaps people needed something to cheer them up and take their minds off their troubles. For whatever reason, science fiction flourished. H. G. Wells's film *Things to Come*, the classic exposition of the technological utopia, appeared in 1935 and did a startlingly good job of predicting the beginnings of World War II and the Blitz. Perhaps aided by the cachet of this apparent foresight, but probably as much by the fact that its iconic designs set the tone for half a century of Art Deco futuristic style, *Things* formed the starting point for most subsequent popular speculation.

The influence of *Things* was followed and extended by the 1939 World's Fair, with its "I Have Seen the Future" pins and its General Motors-sponsored model of the city of the future—the original Futurama. In fact, the phrase "The World of Tomorrow" was coined by Grover Whalen, a principal organizer of the fair.[3] The Futurama exhibit featured the Streamline Moderne designs of Norman Bel Geddes, helping cement them in the public mind as what the future would look like. Geddes's influence was visible in the curved flow of the trains that people rode and the cars they drove, the rounded corners of their radios and home furniture. His dreams of multi-floored passenger airplanes kept company with GM's self-driving car, showcased in 1939 at Futurama.

In *Things to Come*, the cadre of the technological elite that forms the enlightened scientific world government is a community of airmen who call themselves "Wings Over the World." One of the core concepts in Wells's worldview was that a new level of informed, scientific proficiency was being demonstrated by the emerging technological professional class. Airmen were far and away the most visible example, matter-of-factly doing what had been considered impossible short decades before. In Wells's script, the airmen form a cadre of competence that naturally inherits the leadership role after the fall of the self-destructive old order.

Wells's imagined utopia features a technocratic elite rather than a democracy, extremely aggressive deployment of technology, and a Victorian-style exploratory adventurism that values results over individual life. Note that, long after the release of *Things*, many of his concepts were still current enough in the science fiction noosphere to form the background of *Star Trek*.

Wells's heroes, his airmen, are willing to go where none have gone before, and risk everything to accomplish what none have accomplished yet. But, as often happens, Wells's villains are the more engaging—and prophetic. The villains of *Things to Come* are well-to-do people who feel that they have enough for a comfortable life, and don't want any accomplishments going on in their backyard.

As Wells himself points out in the script, it is not a social conflict we are witnessing. The mob trying to stop the film's central moon mission are as well dressed as everyone else. It is not the Haves attacked by the Have-Nots; it is the Doers attacked by the Do-Nots.[4] The Do-Nots are essentially an updated version of the feckless Eloi, the neotenous descendants of humanity from Wells's remarkable first novel, *The Time Machine*. In that story, millennia of comfort and safety have led to degeneration. But, by the time he wrote *Things to Come*, with 40 years more experience in how the world works, Wells produced a more sophisticated and nuanced picture.

## The Postwar World

With World War II, the notion of a technologically driven future that is not merely prosperous but wonderful emerged from science fiction. During the war, any promising technical advance was seized upon and perfected, mass-produced, and put to violent use. Compare the bomber that ushered America into the war, the one-man, one-bomb, sea-level Mitsubishi A6M "Zero" dive-bomber, to the one that saw us out: the nuclear-armed B-29 Superfortress, carrying 10 tons of bombs at an altitude of six miles.

Battleships lost place to aircraft carriers. World War II saw the first use of cruise (V-1) and ballistic (V-2) missiles. The wartime years saw the invention or major development of jets, radar, helicopters, antibiotics, DDT, portable radios, and computers. The aerospace industry quadrupled in size.

Surely, once the war was over all this science, technology, and production could work to the purpose of improving ordinary lives. The writers of the period largely bought into this notion. That had been the premise of *Things to Come* following the First World War, and when the second concluded there was a rapidly growing genre of Wells imitators. This group came to be part of the cluster of ideas that, along with speculation about the political shape of society, came under the heading of "the postwar world."

The kind of thinking that blossomed into "the future we were promised" of the 1950s and '60s can be seen clearly in a 1944 review of two books, *Your World Tomorrow* and *Miracles Ahead! Better Living in the Postwar World*, which ran in *The Saturday Review* under the title "Green Light for the Age of Miracles":

> Half a dozen books have appeared in the last two years in which this recent technological progress is reviewed. The latest two are before us now. They follow the trend of these predecessors.

Of particular interest to us is the description of transportation the books inspired:

> Some of the cars will also have wings so that they can transport their three passengers through the air if need be. Of course there will be helicopter flivvers, and of course we shall speed through the air at low cost in giant airliners, so that we can spend a week-end in London or a vacation in the South Seas or the Himalayas.[5]

Flying cars and helicopter flivvers may not be here, but low-cost giant airliners certainly are. With mild amusement, I note that I've spent a weekend in London and vacationed in the South Seas. But please remember that these kinds of trips were once in the same category as helicopter flivvers. The authors of these books were reasonably well grounded in the technology of the day, and many of their predictions have, in fact, come to pass in the meantime. From the two books in the review, we have:

**Figure 1:** "Remember, Mrs. America likes to go places and see things. And when she finds out that she can cover 600 miles in a morning, to shop or visit in any one of a dozen cities, she's going to fly. And nothing can stop her. ¶ Yes, she'll be flying, taking off and landing, as easily as she drives her car, in a few hours after she first steps in a plane. Compared to the hundreds of American veterans flying today, hundreds of thousands will fly in the post-war tomorrow. Why? Because it's going to be as easy as that for you to fly your own Cessna Family Car of the Air." (Courtesy Cessna.)

> › High-octane gasoline
> › Factory-built housing
> › Portable radios
> › Televisions with screens "the size of a pocket-handkerchief"
> › Air-conditioning
> › Plastics
> › A vast increase in synthetic foods and materials
> › Electronic controls for cooking
> › Fluorescent lights
> › And the aforementioned giant airliners and worldwide vacations

Perhaps the most enduring and popular champion of the "world of tomorrow" throughout the actual postwar period was the avuncular Walt Disney, whose offerings ranged from Tomorrowland at the Magic Kingdom to the planned Experimental Prototype Community of Tomorrow.

We were promised flying cars, and not just in the pages of pulp science fiction. The Cessna ad in Figure 1 appeared in *Flying Magazine* in 1941, and there was a series of similar ones thereafter.

There are similar ads from Piper, Cessna's major competitor of the era, and they continued in this tone well through the postwar tomorrow into the '60s.

Automakers also got into the act. American cars in the 1950s were famously designed to resemble aircraft, with features such as simulated jet intakes and tailfins. Various experiments were done with cars that actually got off the ground, at least a little bit, such as Ford's Levacar (1959) and the Curtiss-Wright Bee (1959).

The bottom line is that it wasn't just science fiction writers telling us that we would have flying cars. The people we expected to make them, and who themselves expected to make them, told us so as well.

The flying car was supposed to be among the more pedestrian achievements of a dawning Space Age. Instead, we now extol the virtues of walking cities built at safe distances from inhospitable airports that use conveyor belts to move bodies from waiting area to waiting area. That was not the space we were expecting to conquer.

Of the 300 covers that graced *Galaxy Science Fiction* magazine from 1950 to 1980, only six featured flying cars. More than half were about space travel.

America's romance with aerospace came to a climax in some sense in 1962, the year of John Glenn's orbital flight. Actual events had overtaken the science fiction of the '50s. Project Mercury Redstones and Atlases eclipsed Tomorrowland's TWA Moonliner.

The Space Age as a phrase and concept had lurked at a very low level in science fiction since the war; in 1952, Random House had added *By Space Ship to the Moon* to its offerings of picture books for children, which had previously consisted of titles such as *Robin Hood*, *Daniel Boone*, *Heidi*, and the like. But the Space Age only lifted off into the public consciousness in 1957, with the appearance of Sputnik.

## A Profile of the Future

Besides Glenn and Project Mercury, 1962 saw the launch of Telstar, the first communications satellite, and the publication of Arthur C. Clarke's *Profiles of the Future*. Best known as one of the world's top science fiction writers (along with Isaac Asimov, Robert A. Heinlein, and A. E. van Vogt), Clarke was also a top science writer and practicing technologist. As the conceptual inventor of the communications satellite, Clarke was, with the 1962 launch of Telstar, at the top of his game, and his predictions of future technology gained extra respectability.

*Profiles of the Future* is an eclectic mixture of takes on future technology, ranging from specific existing machines (hovercraft) to informed speculation on possible capabilities based on new physics (antigravity). Upon publication, it joined a literature of futurism and technological forecasting that included books such as Herman Kahn and Anthony J. Wiener's *The Year 2000*. Half a century later, in 2012, according to Google, each of these books had garnered over half a million references on the web.

It's important to note that Clarke wasn't predicting the immediate future. At one point, he amusedly noted that people trying to solve the problems of the near term would be disappointed that he seemed to be discussing technology that would appear "in the middle of the next century." Indeed, some of his projections were explicitly for the far future. But in the minds of serious thinkers, at least those who bought into technological optimism, *Profiles* was, by all accounts, the definitive take on "the future we were promised."

*Profiles* was subtitled *An Inquiry Into the Limits of the Possible*, but each progression toward those limits started out with a look at technologies that could be reasonably foreseen, in general if not in detail, by a technically informed person of 1962.

A measure of Clarke's astuteness can be seen in the precision with which he nailed a technology that is just emerging 50 years after he wrote: the self-driving car. "The *auto*-mobile of the future," Clark wrote, "will really live up to the first half of its name. You need merely tell it your destination—by dialing a code, or perhaps even verbally—and it will travel there by the most efficient route, after first checking with the highway information system for blockages and traffic jams."[6] He added, "It may one day be a serious offence to drive an automobile on a public highway."

In making such predictions the perspective of a science fiction writer surely helps. Isaac Asimov, in a 1964 article about life in 2014, wrote, "Much effort will be put into the designing of vehicles with 'Robot-brains'—vehicles that can be set for particular destinations and that will then proceed there without interference by the slow reflexes of a human driver."[7] For comparison, Kahn and Wiener's *The Year 2000* (as well as the works of other mainstream futurists, such as Robert Prehoda's *Designing the Future*, a seminal text on technological forecasting) predicted electric cars (along with pneumatic-tube trains), but not self-driving ones. Kahn's closest approach was a listing of "automated highways" as a low-probability possibility. In their defense, this was a perfectly reasonable prediction in 1967. But it was Clarke's genius at inspired guessing just outside the box that made *Profiles* an enduring classic.

Another case in point is Clarke's prediction of worldwide communications networks, where the reality has already overtaken the prognostication:

> We could be in instant contact with each other, wherever we may be, where we can contact our friends anywhere on earth, even if we don't know their actual physical location. It will be possible in that age, perhaps only 50 years from now, for a man to conduct his business from Tahiti or Bali just as well as he could from London. . . . Almost any executive skill, any administrative skill, even any physical skill, could be made independent of distance.[8]

In this respect and many others, Clarke and the others either hit the mark or were not far off. Let's have a look at the things that the major technologically savvy science fiction writers of the 1960s—such as Clarke, Asimov, and Heinlein—predicted for roughly now:

> › Orbital space stations
> › Lunar landings and bases
> › Nuclear rockets
> › Interplanetary travel

› Pocket telephones
› Home-based videophones (also used for reading and commerce)
› Atomic power
› Atomic batteries
› Fusion power
› Space-based military
› Modular machine-built architecture
› Automatic home meal preparation
› Major substitution of synthetic food for agriculture
› Transportation at 1,000 miles per hour and one cent per mile
› Highways in decline as air vehicles predominate
› Self-driving vehicles
› Cyborgs
› Robots (which "will neither be common nor very good in 2014, but they will be in existence"—**Asimov**)
› Automation-eliminating jobs
› Wireless energy transmission
› Translating machines
› Artificial intelligence
› Global library
› Sea mining
› Contraception, revising relations between the sexes
› Cures for cancer, the common cold, and tooth decay
› Psychology and education as hard sciences

Some of these clearly have not happened, and some clearly have, but many are in a gray area. The Nuclear Engine for Rocket Vehicle Application, or NERVA, was built and successfully tested in the '60s, but never flew and the project was canceled.[9] (The Air Force even built and tested a nuclear aircraft engine![10]) The best batteries today are much better than those that 1962 had to offer, but they still fall short of where Clarke (and the others) hoped they would be, i.e., able to replace gasoline in cars. Wireless transmission of energy is similarly advanced but not commercially mature. Artificial intelligence is probably about 80 percent as far along as they thought it would be by now—but Asimov nailed it when it came to robots. Sea-mining technology is here, and it could be a growth industry in the next decades if the legal environment can be worked out.[11] Pocket phones, and the global library in the form of the internet, are here with a vengeance.

As a whole, the science fiction writers were roughly on target, perhaps a decade optimistic on the average, in most fields except transportation and space exploration. It was supposed to be the dawn of the Space Age, and that just didn't happen. The divergence persists with far-future predictions as well: They call for robots in the 2020s, which is almost certainly correct, and for bioengineering in the 2030s, which is very likely conservative—but in transportation they thought we might have interstellar probes, asteroid mining, and gravity control by 2050.

## The Jetsons

Right at the peak of Space Age interest in 1962, Hanna-Barbera introduced *The Jetsons*. From the elementary-age Elroy Jetson's "Little Dipper School" to the dog Astro to the name Jetson itself, everything is predicated on the notion that aeronautics and space travel are a big deal in the animated show's world of 2062.

The Jetsons in the series are a working-class family bearing the same narrative roles as the Kramdens in *The Honeymooners* and the Flintstones in *The Flintstones*. The first episode is something of a tour de force of caricatured *Popular Mechanics* postwar world predictions. We see flying cars, conveyor-belt sidewalks, powered chairs, large-screen 3D TVs, afternoon outings to Acapulco, Rosie the robot maid,[12] automatic sliding doors everywhere, and Elroy going to school via a transparent pneumatic tube.

Everything in the series (including the windows) is designed in swooping curves, with a style reminiscent of the designs of *Things to Come* or Futurama; there are no straight lines. Everything has antennae or tailfins. Clothing is a cartoon mixture of the neoclassical styles of *Things* and 1950s-era skintight pressure suits.

Jane prepares breakfast by pressing buttons on the Foodarackacycle, a menu-like appliance on the counter. The food appears out of an (automatic, sliding) trapdoor in the kitchen table.

Once George arrives at Spacely Space Sprockets, the manufacturing business where he works, his flying car folds up into a briefcase and he glides along a moving sidewalk/conveyor belt from his landing spot into his spacious, windowed office. He then folds his hands behind his head and puts his feet up on his desk (which is actually a control console festooned with dials, buttons, levers, and screens).

It's hard to know where to put *The Jetsons* on a spectrum from pure fantasy to semi-serious, if firmly tongue-in-cheek, forecasting. After all, we would never take *The Flintstones* as a genuine source on what life was like in the Neolithic. The cartoonists simply used a smorgasbord of paleontological creatures, totally out of context, to lampoon current American suburban life. *The Jetsons* was much closer to 1960s reality, but it's clear that the show was also intended to have fun with the *Popular Mechanics*-style world-of-tomorrow optimism.

Did *The Jetsons* simply capture the spirit of the Space Age as reflected in the Apollo program, 50 years of incredible progress in aviation, and the technologically optimistic zeitgeist of the times? I think that they did, in a subtle subtext that is hidden in plain sight by the nature of the show as a cartoon: The Jetsons' lives are ridiculously easy, and they are ridiculously rich. Jane talks to her mother on a wall-size videophone; the show aired during an era when 20 percent of American households didn't have a television and 25 percent didn't have telephones. There is a machine in the Jetson house that does Jane's fabulous hairdo every morning; in the 1960s, a weekly trip to the beauty parlor (complete with iconic beehive hair dryers) was a mark of at least upper-middle-class status.

Mr. Jetson arrives at his office, leans back in his chair, and puts his feet up on the desk. Jane presses three buttons to accomplish the housework—and that's before she gets Rosie. Even the robot sets the table for dinner by pressing a button.

In the postwar era, vacations in Europe and Acapulco were the prerogative of the rich and famous—the jet set—not day trips for schoolkids. The working man's family lived in a cramped row house, not a spacious penthouse. His wife didn't have a house-maid; she spent the whole day cleaning, washing, cooking, and sewing. The man of the house himself worked long, arduous days and was often forced to take public transportation. Private aircraft were a dream almost as far out of reach as they had been before the brothers Wright.

The subtext of *The Jetsons* is simply stated: Advancing technology would make the future much better than the past—for everyone. We could all be jet-setters.

Regardless of its exposition by *Popular Mechanics* or *The Jetsons* or anyone else, the trajectory of history was just such a promise. The remarkable progress of the early 20th century and, indeed, much of the 19th—in particular the rise in material well-being of ordinary people—meant that, if you were an American living in 1960, you could expect that your great-grandchildren would progress as far beyond the way you lived as you had beyond your great-grandparents' way of life.

It's been speculated by levelheaded economists that the decades of remarkable achievement in the early 20th century—of technology, innovation, and economic ex-pansion—were but a set of miracles.[13] It seems miraculous enough, seen with the right historical perspective. But miracles are at once rare, unexpected, and unexplained. The Industrial Revolution had been gathering steam since the early 1700s, more than 200 years before. Low-pressure steam, manufacturing machinery, high-pressure steam, steamboats, and railroads had been possible in some sense since Roman times—remember the inventor and engineer Hero of Alexandria—and of course the opportunities afforded by physics had always been there. But it was the zeitgeist, the spirit of the times, the respect given to Doers instead of Do-Nots, during the Industrial Revolution that made the modern world possible. Physics didn't change. People did.

# 2

# The Graveyard of Dreams

*The fountains are dusty in the graveyard of dreams;*
*The hinges are rusty: They swing with tiny screams.*

—**H. Beam Piper,** ***The Cosmic Computer***

The golden age of science fiction gave us more than just optimistic predictions of technological wonders. Gripping fiction requires conflict. Sometimes the aliens are threatening to invade. Sometimes they have already succeeded. In Philip Francis Nowlan's *Armageddon 2419 A.D.*, the original Buck Rogers story: "I awoke to find the America I knew a total wreck—to find Americans a hunted race in their own land, hiding in the dense forests that covered the shattered and leveled ruins of their once magnificent cities, desperately preserving, and struggling to develop in their secret retreats, the remnants of their culture and science." [14]

In 1960, Detroit, Michigan, was the epitome of American greatness. The automobile was the cornerstone of our economy—some would say our culture. The Beach Boys crooned that she was going to have fun, fun, fun until her daddy took her T-Bird away, reaching number five on the Billboard charts. Motor City had the highest per-capita income in the world.

Fifty-three years later, Detroit filed for Chapter 9 bankruptcy. In the words of the Canadian pundit Mark Steyn:

> By the time Detroit declared bankruptcy, Americans were so inured to the throbbing dirge of Motown's Greatest Hits—40 percent of its streetlamps don't work; 210 of its 317 public parks have been permanently closed; it takes an hour for police to respond to a 9-1-1 call; only a third of its ambulances are drivable; one-third of the city has been abandoned; the local realtor offers houses on sale for a buck and still finds no takers; etc., etc.—Americans were so inured that the formal confirmation of a great city's downfall was greeted with little more than a fatalistic shrug. But it shouldn't be. To achieve this level of devastation, you usually have to be invaded by a foreign power. [15]

Talk about dusty fountains in the graveyard of dreams. In Detroit, the tiny screams of the rusty hinges have swelled to a full-throated cacophony. The prevailing zeitgeist of expectations for the future today is far more pessimistic than it was in 1962, and this has nothing to do with the failure of one or two technological projections to have come to fruition.

Entrepreneur and blogger Paul Graham describes our view of the future with tongue in cheek:

> It's hard to predict what life will be like in a hundred years. There are only a few things we can say with certainty. We know that everyone will drive flying cars, that zoning laws will be relaxed to allow buildings hundreds of stories tall, that it will be dark most of the time, and that women will all be trained in the martial arts.[16]

He's referring to the now-standard film noir depiction of the future city as seen in films such as *Blade Runner*: a world dark, hard, and artificial, where life is nasty, brutish, and short.

Expectations of flying cars backed up by the deliberative efforts of engineers, manufacturers, and financiers were the optimistic stuff of the 1930s, '40s, and '50s. Pause over that. During a Great Depression, a World War, and the start of a globe-spanning Cold War, Americans were not only promised flying cars, but they could read news stories of steady progress toward the flying cars' production. But as we slipped into the 1960s and '70s, more and more Americans began to lower their sights toward a less hopeful expectation. Wilting optimism accompanied stagnating growth.

"Heeding the warnings of these forecasting missteps," Robert J. Gordon wrote, "we should also recall the past overoptimism, including the universal prediction in the late 1940s that within a generation each family would have its own vertical lift-off airplane, a universal society of Jetsons."[17] Or Tyler Cowen's succinct comment: "We're very far from the flying car."[18]

To many people of 1962, including my eight-year-old self, it was inconceivable that we would be using the same flying machines now that we did then. "What will our transportation technology look like in 50 years?" you might have asked them. And they would have answered, "Well, look what it was like 50 years ago, in 1912!"

In 1912, the vast majority of humanity had never even seen a flying machine. The brothers Wright had only just begun selling the Wright Model B, the first production airplane. (See Figure 2.) Louis Blériot had only recently crossed the English Channel (Calais to Dover, about 25 miles, at 40 knots) in a machine that looked like a kite tied to a tricycle with baling wire—and caused a worldwide sensation. The Fokkers and Sopwith Camels of the World War I flying aces were still years in the future.

By 1962, fighter planes such as Lockheed's F-104 Starfighter flew at over 1,300 miles per hour. A bomber such as the B-52, which had been in production for 10 years, had a loaded weight of over 250,000 pounds and a range of over 10,000 miles. Even the commercial airliners of the day, such as the 727 and the DC-8, could carry 170 people 5,000 miles at 600 miles per hour.

Over the same period, the automobile, which had been a toy for the rich in 1912, had become universally owned. Indeed, it had become so ubiquitous that it was a cor-

**Figure 2:**  Fifty years, 1912 (top) to 1962 (bottom).

nerstone of the American way of life. If the Rolls-Royces, Packards, and Duesenbergs of the 1910s could pave the way to the Fords and Chevys for everyone 50 years later, surely the Beechcrafts, Mooneys, and Sikorskys of the '60s could lead to Pipers, Cessnas, and Hillers for all today. Or, rather, to something as far beyond the Pipers, Cessnas, and Hillers of the '60s as they themselves were beyond the Wright Model B or the Blériot XI.

People in 1962 had a very reasonable basis for expecting technological progress to improve their lives a lot in the coming decades. It had done so substantially in the previous decades. The life of the average American in 1962 was enormously better than it had been in 1900. Obvious changes were the automobile itself, flying machines, skyscrapers, antibiotics, movies, premade clothing, electric lighting and power in the home, radio, television, refrigerators, vacuum cleaners, washing machines, and stoves that turned on with a touch of a button instead of your having to build a fire in them—after having split the wood by hand.

Furthermore, people could afford to buy the new machines and products. The average American had become three to four times richer between 1900 and 1962.

**Figure 3:** You call this the future? (Calvin and Hobbes © 1989 Watterson. Reprinted with permission of UNIVERSAL UCLICK. All rights reserved.)

GDP per capita grew from $4,000 to $5,000 (constant 2005 dollars) in 1900 to $16,000 in 1962. Personal income as a fraction of total productivity remained consistent at roughly 50 percent throughout the period. Behind the scenes, a host of innovations, from assembly-line manufacturing to containerized cargo shipping, boosted productivity. Luxuries previously reserved for the most wealthy began to come within the reach of most everyone.

Surely it was reasonable to imagine that the same kind of progress that had made the automobile, a rich man's plaything in 1910, into an indispensable adjunct of the average family by the mid-20th century could do the same with the private aircraft by the mid-21st.

Surely it was reasonable to imagine that flying machines themselves, which the scientific community firmly believed impossible in 1900, but which had reached supersonic speed in 1947 (while Orville Wright was still alive), and which were carrying passengers in intercontinental jets in 1962, might carry humans to other planets by 2062.

Surely it was reasonable to imagine that, following advances in medicine that added three decades to the average lifespan, further advances, possibly even rejuvenation, were in prospect.

## Stagnation Point

"You call this the future?" Calvin (the little boy) famously complains to Hobbes (the imaginary tiger) in the cartoon. (See Figure 3.) "Where are the flying cars?" That was December 30, 1989. A third of a century has gone by since then, and we're still waiting. The question "Where is my flying car?," Wikipedia tells us, "is emblematic of the supposed failure of modern technology to match futuristic visions that were promoted in earlier decades."[19] A Google search for the question returns 781,000 webpages. Flying cars have become a symbol of a mismatch: The future as imagined in the first half of the 20th century seemed a lot brighter than the present we're living in now.

We have seen the future, and it doesn't work as well as we expected. Progress has slowed to a crawl. Or is that merely perception?

After all, in 1989 there was no Wikipedia and no Google. Indeed, there were no webpages. The World Wide Web would make its public appearance, at the CERN laboratory in Europe, in 1991. The internet itself connected mostly researchers at universities and government labs. As cyberpunk author William Gibson observed, "The future is already here—it's just not very evenly distributed."

Adam Gurri, blogging at The Ümlaut, wrote:

> We wanted flying cars, instead we got the ability to instantly connect with anyone anywhere in the world, to share stories, pictures, music, podcasts, ideas, film, animation, comics, feedback, friendship, love, and our lives. Flying cars seem really cool on their face, but I somehow doubt that they would have so meaningful an impact on our lives.[20]

After all, George Jetson didn't even have a cellphone.

So, did the prognosticators of a half-century ago merely miss the mark? Kevin Drum, blogger at *Mother Jones*, thinks so:

> The argument here is that back in the 1950s we thought the future would bring us flying cars, electricity too cheap to meter, and vacations on the moon. But none of that has happened. What gives?

> The answer is prosaic: Forecasters in the 50s were wrong. It's not that the future never arrived—it's that the future brought us different stuff than we thought we were going to get. Our lack of flying cars simply doesn't tell us anything about the pace of innovation.[21]

Drum was writing in reaction to a number of recent commentators claiming that meaningful innovation had slowed to a crawl in the latter 20th century. Many people (including me, in the book *Nanofuture*) have pointed out that the second half of the 20th century didn't seem to bring nearly as many major, valuable advances as the first half. Instead, it brought constant distraction and annoying clickbait. Entrepreneur Peter Thiel has written extensively about "The End of the Future," but he sums it up masterfully in a short quip: "We wanted flying cars; we got 140 characters."[22]

Clearly, there's more progress in some ways than expected, to match the less progress that happened in such areas as flying cars. Yet even this was predicted: Arthur C. Clarke, in *Profiles of the Future*, offers one of the best explanations from a purely technical point of view, about why progress in fast transport might fall short of predictions:

> There is, however, one trend which may work against the establishment of a virtually instantaneously global transportation system. As communications improve, until all the senses—and not merely vision and hearing—can be projected anywhere on the face of the Earth, men will have less and less incentive to travel. This situation was envisioned half a century ago by E. M. Forster in his famous short story "The Machine Stops,"[23] where he pictured our remote de-

scendants as living in single cells, scarcely ever leaving them but being able to establish instant TV contact with anyone else on Earth, wherever he might be.

In his own lifetime Forster has seen TV perfected far beyond his imaginings of three decades ago, and his vision of the future may be, in its essentials, not so far from the truth. Telecommunication and transportation are opposing forces, which so far have always struck a balance. If the first should ever win, the world of Forster's story would be the result. [24]

This ranks among Clarke's most perceptive predictions. What he says about the effects of technology on the average person is quite true. *The Leave It to Beaver* family of the '50s had a life that was essentially as comfortable as ours is today, but worlds removed from that of the turn of the 20th century. Improvements that we've enjoyed since Ward and June Cleaver's day, however, as seen by the average person, have consisted to a remarkably large extent of ever more sophisticated, and ever more intrusive, entertainment. Oh yes, and advertising.

This is the exact trajectory of decline leading to Wells's Eloi. But, of course, the Eloi don't see it that way; the Do-Nots favor stagnation and are happy turning our civilization into a collective couch potato. They will tell us that the promised future is not simply late, but that it was never realistic to begin. But we know better.

It is nearly 50 years since I sat glued to a grainy black-and-white TV set and watched Neil Armstrong and Buzz Aldrin land on, and then step out onto, the moon. If you had asked me then, I would have blithely assured you that by the year 2000, much less 2020, I'd have my own spaceship, or at least own a flying car and be able to buy tickets on a spaceship to the big celebration at Tranquility Base.

It didn't work out that way. Not only am I not going to be celebrating on the moon, nobody is going there for any purpose whatsoever. Only 12 men have ever walked on the moon. All of them were born while Orville Wright still drew breath and no one had flown a jet. Only Aldrin, Duke, Schmitt, and Scott are still alive as of this writing; Armstrong, Bean, Cernan, Conrad, Irwin, Mitchell, Shepard, and Young are gone. John Glenn, last of the original seven Project Mercury astronauts, died in 2016 at the age of 95. Godspeed. But with their passing we mourn more than the age of heroic astronauts. We mourn a type of hero, as captured in the words "I am and ever will be, a white-socks, pocket-protector, nerdy engineer.... I take a substantial amount of pride in the accomplishments of my profession." [25] Thus spoke Neil Armstrong, whose actual accomplishments, capped off by the famous small step, should have engendered some giant leaps in the meantime.

## Flatline

A lot of the recent commentary about the slowing or not of technological progress was prompted by the 2011 appearance of *The Great Stagnation* by George Mason University economist Tyler Cowen, a book likely prompted in part by some observa-

**Figure 4:** Data from Turchin, Ages of Discord, figure 3.3. The smooth line represents a steady 2 percent growth rate for comparison.

tions of Peter Thiel. Cowen's thesis is that, following a period of remarkable progress due to humanity's ability to easily pick low-hanging technological fruit, the United States has entered a slumping economy, a fact apparent since the 1970s. This is reflected in the slow growth of income and the loss of the American Dream, the faith that your children will have a better life than you did.

Cowen uses median income rates to argue for the Great Stagnation, but we can also see it in this flatline in wages of unskilled labor, from statistical historian Peter Turchin. (See Figure 4.)

This is perhaps the most striking example, but Turchin points out that the entire set of trends he tracks shifted from a favorable ("integrative phase") direction to an unfavorable ("disintegrative phase") one in the 1970s.

Even worse is a phenomenon that some economists (including Cowen) have taken to calling "cost disease," which is infecting more and more of modern life. There are some things—typically the most important things—whose costs keep stubbornly going up in real terms, that is, even after being adjusted for inflation. Some examples: Housing costs twice as much, on average, and primary education costs three times as much as in the '60s, and children are not learning more. Until the '70s, health care costs and longevity in the United States grew at about the same rates as in comparable developed countries; since then, longevity has grown more slowly and costs have grown much faster. Medical care now costs six times as much as it did in the '60s: In 1960, the average worker's yearly health insurance cost 10 days' salary; today, it costs 60 days' worth.[26] College tuition and textbooks cost in the neighborhood of 10 times as much. Note that this is entirely different from the well-understood phenomenon in which some things seem to get more expensive because others are more efficiently supplied, for example, by mass production. Given the current level of technology, for example, a textbook could cost $1 instead of $100.

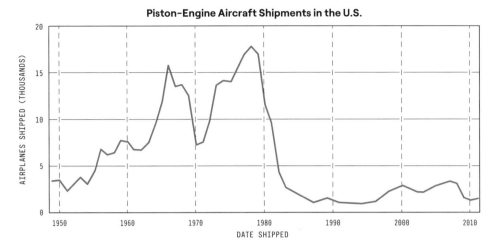

**Figure 5:** The private airplane industry crashed around 1980.

In the '70s, actual economic growth fell below a constant-growth trend line and never recovered. We went into a secular decline that has us farther from the curve now (in dollar terms) than we were in the depths of the Great Depression itself. In the four decades prior to 1970, the per capita GDP had a growth rate of 2.8 percent; in the four decades after, 1.9 percent. In the postwar era (i.e., 1945-1975), we stayed quite close to a constant growth-rate curve. If you were a futurist in the '60s, you could be forgiven for assuming that we had the economic know-how capable of sustaining that growth—after all, we had demonstrated it for a quarter-century. In fact, futurists in the '60s believed exactly that:

> The most significant discovery emerging from our increasingly sophisticated understanding of the economic growth process is the fact that accelerated economic growth can be achieved by a combination of fiscal policy, monetary policy, and continuously increasing R&D budgets (both Federal and private).[27]

That was Robert Prehoda writing in *Designing the Future*, in 1967. A 1966 book entitled *The Shape of Automation*, by Herbert Simon, one of the founding fathers of artificial intelligence who would go on to win the Nobel Prize in Economics, said essentially the same thing: that, because of the increasing productivity due to automation, the average family income would reach $28,000 (in 1966 dollars) after the turn of the century. (An easy rule of thumb for translating 1960s dollars to current-day ones is to multiply by 10.)

Note that Simon's projection was only for a 3 percent growth rate. This was not a ridiculous assumption; real family income had grown at roughly 2.5 percent annually during the previous 50 years. If "our increasingly sophisticated understanding of the economic growth process" had been real, it should have been easy to bump it up another half-percent. But what actually happened in the succeeding half-century is that the family income growth rate went from 2.5 percent to 0.5 percent.[28]

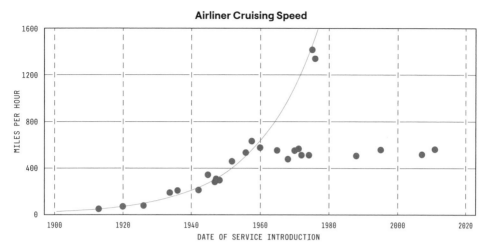

**Figure 6:** Airliner speed vs. date. The curve is a simple exponential fit to points before 1960. Data from Wikipedia.

Something went wrong, though it isn't clear what. For example, in that succeeding half-century the number of single-parent households rose much faster than the population as a whole; lower family income doesn't necessarily mean lower per capita income. But it does mean that many of the economic efficiencies that a traditional family structure provided have been lost.

Any one of these indicators of a stagnation might readily be explained away as reflecting a different trend, or as a misunderstanding of the statistics by your admittedly non-economist author. But it is difficult, I think, to dismiss all of them.

## The Airplane Crash

Another thing that happened right around the end of the '70s was that the private airplane industry, including companies such as Cessna and Piper, which had both been founded in 1927, mysteriously disappeared. (See Figure 6.)

The 1960s had been a boom time for private aviation; Jules Bergman's 1964 book *Anyone Can Fly* has a 40-page appendix displaying the various makes of small family-size airplanes available to private owners.

In Figure 5, which shows the number of private (piston-engine propeller) airplanes shipped, you can see the effect of the energy crisis as the dip around 1970, but it's also evident that the industry had bounced back from that in half a decade. It was perfectly reasonable to have expected, in the '60s and '70s, a continued increase in the numbers, quality, performance, and capabilities of private airplanes. But that didn't happen.

Today, only about 700 new piston airplanes are sold each year, and about 400 turboprops and 400 private jets. Given such limited demand, airplanes cost more than 10 times what they did in the '70s. There are other, less obvious falterings as well. For a graphic demonstration, you can't do much better than looking at airliner speeds. They rose on a nice exponential growth curve until the '60s, and then flat-lined. (See Figure 6.)

This particular plateau can be explained easily enough by physics and economics: It's three times as expensive to fly just above Mach 1 as just below it. The Concorde SST was heavily subsidized as a national prestige project by Great Britain and France (as was the USSR's similar Tupolev Tu-144—see the two points at the top of the chart). As a measure of what the technology was capable of, though, those two aircraft fell right on the trend line, as you can see. For example, the de Havilland Ghost turbojet engine in the '50s had 5,000 pounds of thrust; the GE90, which entered service in the '90s, has 115,000.

The people of 1962 had seen a solid order of magnitude increase in the speed and in the range and in the carrying capacity of available aircraft over the previous 50 years. Another 50 should have brought another order of magnitude. Without the flatline, the trend curve would put us at over 10,000 miles per hour—New York to Sydney in an hour.

Don't forget the other figures of merit, either. Airliners should have been getting more capacious and comfortable at the same time that they were getting faster, as they had been since the days of the Ford Tri-Motor.

In the 50 years since the flatline, however, we've seen virtually nothing; indeed, in a bid for fuel efficiency, the average airliner today is somewhat slower (and more cramped). And yet, given the rise of a substantial trans-Pacific trade in those 50 years, it seems almost certain that there is a market for cutting, say, six hours off the LA-to-Tokyo travel time. The technology to do so certainly exists. It turns out that, for long international flights (e.g., Los Angeles to Sydney), the energy used by a jetliner dragging its way through the atmosphere is just as much as would have been needed to put it into orbit—and make the trip in less than an hour. With an easy-to-handle acceleration, 1 G for five minutes and the reverse to slow down, you could get from New York to San Francisco in 22 minutes.

So, while there is an economic "rut" in the design space for airliners, the fact that we have been stuck in it for half a century remains a good indication that there is some other roadblock in the way of technological improvement. We can reasonably use the plateau in air travel as an iconic, if perhaps exaggerated, example of the general plateauing, flatlining, and stagnating of life-improving technology that began in the 1970s and '80s. A look back at another major advance in long-distance travel technology is instructive. Wooden sailboats were much cheaper to build than steel steamships, and used no coal at all. But the extra expense was well worth the result.

## The Henry Adams Curve

Henry Adams, scion of the house of the two eponymous presidents, wrote in his autobiography about a century ago:

> The coal-output of the world, speaking roughly, doubled every ten years between 1840 and 1900, in the form of utilized power, for the ton of coal yielded three or four times as much power in 1900 as in 1840. Rapid as this rate of ac-

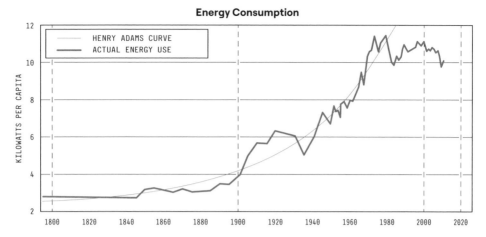

**Figure 7:** Energy consumption per capita in the U.S. One kilowatt, of course, equals 8,766 kilowatt-hours (kWh) per year.

celeration in volume seems, it may be tested in a thousand ways without greatly reducing it. Perhaps the ocean steamer is nearest unity and easiest to measure, for any one might hire, in 1905, for a small sum of money, the use of 30,000 steam-horsepower to cross the ocean, and by halving this figure every ten years, he got back to 234 horsepower for 1835, which was accuracy enough for his purposes.[29]

In other words, going back at least to the Newcomen and Savery steam engines of 300 years ago, we have had a very steady trend of about 7 percent yearly growth in usable energy available to our civilization. Let us call it the "Henry Adams Curve." The optimism and constant improvement of life in the 19th century and the first half of the 20th can quite readily be seen as predicated on this curve. To a first approximation, it can be factored into a 3 percent population growth rate, a 2 percent energy-efficiency growth rate, and a 2 percent growth in actual energy consumed per capita.

The Henry Adams Curve, the centuries-long historical trend, can be rendered as the smooth red line. Since the scale is power per capita, this is only the 2 percent component. The blue curve is actual energy use in the U.S., which up to the 1970s matched the trend quite well. But then energy consumption flatlined.[30] (See Figure 7.)

The 1970s were famously the time of the OPEC oil embargo and the "energy crisis." But major shortages preceded the embargo by a year or two. They were caused by Nixon's energy price controls, instituted in 1971. The embargo didn't occur until 1973.

If you didn't know better, you would think that the Department of Energy was established, on August 4, 1977, with the intent to prevent energy use.

Before the Industrial Revolution, most of the power that humans used was produced by their own muscles, or those of domesticated animals. A human eating a diet of 2,000 calories per day can produce a daily average of about 100 watts. The average South American today uses 10 times that amount of power, and the average North American 100 times. It doesn't seem at all unreasonable to imagine another

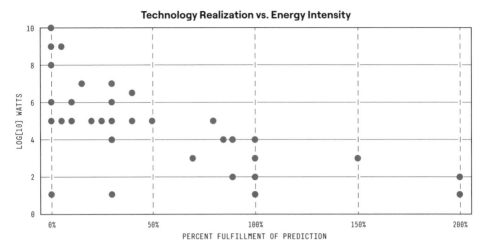

**Figure 8:** Technologies predicted in science fiction ca. 1960 (see Chapter 1), with percent of fulfillment today vs. energy intensity. Details in the Appendix.

Industrial Revolution giving us another factor of 10, or even another 100. We would already be up an extra factor of two or three by now if we had only continued the energy trajectory that we were on before 1970. And we might just have had flying cars, space travel, and the rest before 2062. But the Henry Adams Curve, the heartbeat of our civilization, flatlined.

This had real and continuing consequences. There has been a marked drop-off in the technological advances that make a big difference in people's lives—measured in productivity, health, and, yes, speed and ease of getting around. We've already seen that the list of life-changing technologies that revolutionized the early 20th century, up to and including the automobile itself, was largely not continued in the latter half. The authors of the golden age of science fiction form something of a canary in the coal mine. Their technological predictions that came true, and the ones that didn't, form a striking pattern. (See Figure 8.)

On the power (vertical) scale, our current energy flatline of 10 kilowatts is a 4; below that level, technologies have fulfillment levels all across the bottom of the chart, including some more successful than the predictions, with a nice concentration centered around 100 percent.

Similarly, over on the left side of the chart, predictions run all the way up and down the power scale, with a cluster at 5 (e.g., 100kW), which would have been roughly right for now without the flatline. But all of those have fulfillment levels of less than 50 percent.

But the top right seems to be completely forbidden. The extent to which a technology didn't live up to its *Jetsons*-era expectation is strongly correlated with its energy intensity. The one area where progress continued most robustly—Moore's Law in computing and communications—was the one where energy was not a major concern. It was my privilege—blind luck, really—to have spent most of my working scientific life in that area, and so I arrived at the 21st century with some sense of how it should have been—indeed, *how it had been*—for everyone in the technological enterprise up until my benighted generation.

"Correlated" actually doesn't begin to tell the story. When I began gathering the data for that graph, I had expected a correlation, a general drift or trend where a greater need for power predicted a smaller chance of success. But the result took me by surprise: It looks like someone took a machete and whacked off the top right half of the graph in one fell swoop. That, the top right half, represents the futures we were promised but were denied. Including flying cars.

> *Power is our only lack. We generate all we can with the materials and knowledge at our disposal, but we never have enough. Our development is hindered, our birth-rate must be held down to a minimum, many new cities which we need cannot be built and many new projects cannot be started, all for lack of power.*

—E. E. "Doc" Smith, *Skylark Three*

# 3

# The Conquest of the Air

*Few people, I fancy, who know the work of Langley, Lilienthal, Pilcher, Maxim, and Chanute, but will be inclined to believe that long before the year AD 2000, and very probably before 1950, a successful aeroplane will have soared and come home safe and sound.*

—H. G. Wells, *Anticipations*

Circa 1900, it was the received wisdom of the scientific community that powered, heavier-than-air flying machines were impossible, or, as *The New York Times* noted in 1903, at least a million years off. There were, of course, no real experts on flying machines to consult, but the scientists that the newspapers liked to quote as authorities on the subject were people such as Simon Newcomb, who was an astronomer. I suppose that he did have some experience looking up into the sky. His reputation today rests almost entirely on the various critiques he offered against the possibility of flying machines.

But it is less than satisfying to chuckle at Newcomb's expense, only to flag a taxi to the airport where you wait for an hour or more to board an airplane. Yes, you have flown—to another airport where, if you're lucky, within an hour of arriving you'll have been given keys to a rented car. This sequence is so familiar that you take the hurry-up-and-wait for granted. Newcomb and the scientific world, taking different things for granted, were to learn within his lifetime that his conclusions were wrong. But many of his individual arguments were cogent. They pointed out problems to be solved rather than showstoppers that precluded flight.

For example, in 1903 Newcomb opined, "The example of the bird does not prove that man can fly. Imagine the proud possessor of the aeroplane darting through the air at a speed of several hundred feet per second. It is the speed alone that sustains him. How is he ever going to stop?"[31]

How indeed does an airplane stop? In fact, we currently manage it by a kludge that is one of the least satisfactory parts of our flying technology: The airplane descends to the ground at flying speed, and only slows down after it has rolled along a lengthy runway. If it were not for this necessity, any airplane would be a perfectly usable flying car.

## The Power Curve

The key to flying is energy management. The airplane needs power to stay up (generate lift) and to maintain speed (overcome drag). That power comes from three

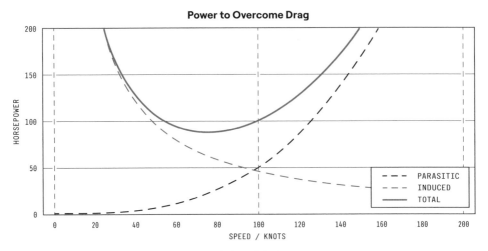

**Figure 9:** Power curve for a fictitious but not unrealistic small airplane.

stores of energy: You have enough energy in your speed to keep flying a few seconds, in your altitude for a few minutes, and in your fuel tank for a few hours. Your takeoff roll tops up your speed in a few seconds; you spend a few minutes climbing to altitude. As long as your engine is working, this is straightforward.

Which brings us to Simon Newcomb's point. Indeed, the problem is worse than he realized. The problematic part of flying is landing: If you simply point the nose of your airplane downward, you will accelerate alarmingly. At just the time you wanted to slow down, your altitude energy is converted to speed. So, instead, you take a long, slanting approach trajectory for landing, not unlike takeoff and climb, to let the energy from both altitude and speed bleed off. But, even so, you have to be moving at flying speed all the way down, slowing only after your wheels have touched the ground.

Induced drag, associated with the power necessary to create lift, decreases with increasing speed. Parasitic drag, associated with all the other useless stirring up of the air in a plane's wake, increases. Add these drags together for the total power necessary to fly at a given speed. The resulting graph is called the power curve. (See Figure 9.)

Intuitively, flying faster than your optimal speed requires more power. Unintuitively, flying *slower* than your optimal speed requires more power, too.

Go to Trafalgar Square and watch the pigeons. To land, they spread their wings wide in a high-lift, high-drag configuration and soar in toward their target spot. Then, in the last half-second, they flap vigorously. As the pigeon slows down to land, it moves into the left side of the power curve and uses more power than it would to fly fast.

This is the one thing an airplane of classical design cannot do, and not simply because the wings don't flap. The fixed-wing airplane is essentially a glider with a built-in towing system. In general, to use the propeller to increase lift also increases speed. Put bluntly, right when you are trying to slow down, you are speeding up to increase lift. But birds, because their method of producing thrust is integrated with the lift,

Figure 10:  Zimmerman's Flying Pancake (the Vought V-173), an airplane suited to the needs of the private owner.

can apply full power, increase lift, and slow down all at the same time. This is why birds land on branches instead of on runways.

The closer the technology of flight could come to doing what birds do, the closer we'd be to flying cars that could land on a short driveway. But we became so accustomed to a stagnant aeronautical industry over the second half of the 20th century that we have forgotten how fast it improved during the first half—and how close we came to exactly that.

The original heyday of the flying car was the 1930s, a fact almost completely forgotten today. Charles Zimmerman, an engineer at the National Advisory Committee for Aeronautics (NACA, revamped and renamed NASA in 1958), developed a "Flying Pancake"—a low-aspect-ratio flying wing—in that long-ago decade.[32] (See Figure 10.) Today, it is mostly remembered for being investigated by the Navy for aircraft-carrier use during World War II. But Zimmerman's 1932 study begins:

> In recent years there has been an increasing demand for an airplane suited to the needs of the private owner. Without going into a discussion of the problem it may be said that such an airplane should be capable of descending along a steep path at such a low rate of speed that it will be unnecessary for the pilot to alter the direction of the flight path or the speed when near the ground in order to make a satisfactory landing.[33]

Zimmerman was on to the problem of the runway.

The runway at the small county airport near my home is a full 12 acres of concrete; the airport as a whole is 100 acres. This isn't going to fit on your suburban lot, or even your country estate. Zimmerman was attempting to design a plane that could emulate the pigeon.

**Figure 11:** W. R. Custer and his Channel Wing.

There are various examples in aviation history of attempts to do the same thing that Zimmerman did with his Flying Pancake, or at least approach it. The Custer Channel Wing aircraft, for example, illustrates a technique called "upper-surface blowing." (See Figure 11.) Willard R. Custer had patented his Channel Wing in 1929, although it didn't fly until the '40s. Briefly pause over that fact. This craft flew in the 1940s, and the pictured plane, the CCW-5, is reasonably close to the specs of the air-car that we would like to have today. It seated five passengers, weighed two and a half tons, and cruised at 200 miles per hour. It could also fly at 35 miles per hour and had a 250-foot takeoff roll. That's the shadow of a parked Boeing 747-8.[34] The Custer achieved its low-speed lift by positioning the propellers so that the intake caused a high-speed airflow over the upper surface (but not the lower surface) of the semicircular wing section. The main problem with the CCW, which also afflicted the Flying Pancake, was that you had to tilt the entire vehicle to an extremely high angle of attack to get the low-and-slow characteristics. (Note that the pigeon does the same thing.) This was not only a bit more like a carnival ride than your average passenger bargained for, but the position obscured ground visibility just when it was needed most.

The pigeon, the Pancake, and the Custer all use a lot of power when flying slowly, just as the power curve demands. But, at the same time, the landing aircraft is trying to get rid of excess energy in the form of altitude and speed. What if we could use this energy for the power we need to fly slowly?

## Autogyro

Juan de la Cierva was one of the great early aviation pioneers. He built Spain's first airplane, but then one of his early models stalled due to moving too slowly and crashed. De la Cierva set himself to find an improved design. He reasoned that if he could make the wings move fast even when the craft itself was moving slow, he might solve the stall problem.

In the early 1920s, de la Cierva invented the autogyro, which used what we would now call a set of rotor blades for a wing. The rotor isn't powered; it is, in fact, a rotary wing. Because it rotates, the speed of the blades through the air can be higher than that of the overall craft, but the dynamics are those of an airplane rather than a helicopter: The wings create lift but cause drag; there must be a separate propeller to provide thrust to overcome the drag. In simplest terms, the wind passing through the rotor disc causes the rotors to turn ("autorotate"), just as wind causes the arms of a windmill to turn. Each rotor blade is essentially gliding "down" into the oncoming wind.

There were several innovations necessary to produce a working autogyro. The first problem to overcome was dissymmetry of lift. The rotor turns such that the blades on the right side of the aircraft are moving forward while those on the left side are moving backward. So, when the aircraft itself moves forward, the right side generates more lift due to the added speed, and the left loses lift the same way. Thus the rotor will try to flip the aircraft over onto its left side.

De la Cierva solved this in a typically ingenious way: He put in a hinge that allowed the blade to flap up and down as it went around. This is far from obvious, and in fact the solution is both subtle and complex. But you can get an intuitive grasp on it by thinking about the flapping wings of a bird. It's not so hard to imagine that the bird's wings produce less lift when flapping upward than when flapping downward.

There is another reason a flapping hinge is counterintuitive: How can a wing free to flap upward hold up the airplane? At least the answer to this question is a bit more obvious: Centrifugal force holds it out. The gyro, and indeed a modern helicopter, hang from their rotors the way that a weight would hang from the middle of a clothesline.

We are privileged to understand many of de la Cierva's thought processes as he solved the challenges of the autogyro. He explained them in his very readable autobiography, *Wings of Tomorrow*. He also speculates there that a workable helicopter may never be built, not realizing that he himself was solving most of the problems standing in its way.

It took most of the 1920s to develop the autogyro into a useful and reliable form.

For much of this we have de la Cierva, a legitimate and recognized aeronautical genius, to thank. But he had plenty of help, and many loyal followers. For example, de la Cierva couldn't see any aerodynamic reason for the necessity of lead-lag hinges in the hub to allow for the acceleration and deceleration of the blades as they went around. He argued with test pilot Frank Courtney that they shouldn't be necessary. Courtney won the argument when, due to metal fatigue, two blades broke off an autogyro that he was flying in 1927. (Luckily, he survived.)[35]

Ironically, the acceleration of the blades is due to the flapping hinges, de la Cierva's signature invention. Flapping causes the blades to emulate a twirling ice-skater's arms. Pull them in, and the skater's spin accelerates. De la Cierva's flapping blades effectively went in and out, just enough to matter, with every rotation. Not for the last time it was the man in the machine, not the one in the office, who noticed the problem with the urgency that it deserved.

De la Cierva refined the autogyro through the 1920s, making and selling them in England. In 1929, he was visited by American aviation pioneer Harold Pitcairn, who had built the successful aircraft company that ultimately became Eastern Airlines. Pitcairn was quite taken with the autogyro, bought the rights to build and develop it in the United States, and cross-licensed and collaborated with de la Cierva in its further development. In 1931, a Pitcairn autogyro landed on the White House lawn—at President Hoover's suggestion—as part of the ceremony in which Hoover presented Pitcairn with the Collier Trophy for outstanding achievement in aviation. The same year, Amelia Earhart made a transcontinental flight and set an altitude record in an autogyro; she published an article in *Cosmopolitan* in August, saying that soon "country houses would have wind cones flying from their roofs to guide guests to the front lawn landing area."[36]

By the 1930s, autogyros had captured the public's interest. In fiction, everyone from W. C. Fields to Doc Savage flew one. The 1934 movie *It Happened One Night* showed a groom arriving at his wedding in an autogyro, landing on the lawn. In 1935, *Things to Come* featured an autogyro-like craft, sporting a Norman Bel Geddes Streamline Moderne style. Robert Heinlein's flying car as described in the novel *For Us, the Living* (1938) is quite similar. This was before the need for a side-facing tail rotor in a powered-rotor helicopter was understood, so the fictional designs looked like the working autogyros of the day, with thrust propellers.

The gyros of the early '30s had wings, but the ones later in the decade did not. The difference was what was called the "direct control" rotor hub, invented by de la Cierva and incorporated on the C-30 and most later Cierva and Pitcairn models. This meant that the pilot had a direct handle on the cyclic pitch of the rotor, rather than influencing it by way of the attitude of the fuselage (as by ailerons on wings).

In 1936, in one of the ironically tragic knife twists of history, Juan de la Cierva was killed on his way back to Spain when a conventional commercial airliner (a DC-2) crashed on takeoff in Croydon, England.[37] Besides the personal tragedy, this was a disaster for the autogyro community. De la Cierva had been not only its intellectual but, to a large extent, its spiritual leader. After his death, one can sense in the historical record a breakdown in the friendly rivalries among the various inventors, engineers, and entrepreneurs, and the rise of real antipathies.

The English company that de la Cierva had founded quit autogyro work to focus on helicopters, refusing to share technical information with Pitcairn, even though they were contractually obligated to do so. Somewhat inexplicably, given the looming hostilities of World War II, they licensed their technology to the German firm Focke-Wulf, which flew the first practical helicopter, the FW-61, in 1936. German submarines carried the Focke-Achgelis FA-330, a towed observation gyrocopter on a 400-foot tether, during the war.

Back in the U.S., Pitcairn continued with refinements of his company's autogyro, going through some 36 models during the 1930s. Ultimately, de la Cierva and Pitcairn invented almost everything that goes into a working helicopter.

Figure 12:   The Pitcairn AC-35 roadable autogyro on the streets of Washington, D.C. (National Air and Space Museum).

If it hadn't been for the Great Depression and then World War II, the autogyro might well have gone on to be the basis for a widely used flying car. Its STOVL (short takeoff, vertical landing) capabilities made it amenable to landing on and taking off from driveways or parking lots. The AC-35 could fold its rotors, drive on the streets, and park in an ordinary garage. (See Figure 12.) And it had in Pitcairn an experienced, competent, well-funded sponsor who was committed to the goal of private flying machines.

While homebuilt ultralights piloted (and even sometimes designed) by their build-ers are prone to accidents, properly designed gyroplanes with properly trained pilots have a number of safety advantages, starting with low takeoff and landing speeds. They are also less susceptible to gusty wind than are airplanes: The high speed of the rotor makes the wind speed a smaller factor by comparison. The gyroplane also has some safety advantages over the helicopter; the helicopter can be shifted into autorotation in the case of an engine failure, but there are parts of the flight envelope where this is im-possible to pull off in time. The autogyro is already in autorotation: In one of its very ear-liest flights, de la Cierva's first working autogyro, the C.4, suffered an engine failure on takeoff at about 30 feet altitude. This is a *very* dangerous situation in an airplane: You're likely to stall, you're typically out of runway, and you have no room to maneuver. But the C.4 simply settled to a safe landing. Gyros can land on a dime with the engine out—I've done it myself, in practice, many times. Over either mountains or cities, this is a distinct advantage, because a nice, long, flat place to land is often hard to find!

Takeoff in a gyro is more difficult than in a fixed-wing airplane, but landing is easier. The ratio of gyro accidents on takeoff to accidents on landing during the 1930s bears this out (as does my own personal experience). The most common accident in a plane is slipping, stalling, and crashing on approach. In a gyro, it is tipping over on takeoff, causing a rotor ground strike. You are much more likely to walk away from a tip-over, where you are already on the ground, than from an uncontrolled fall from 50 stories up. And even the gyro takeoff can be simplified by a more powerful pre-rotator, which allows you to skip the hairiest phase of the takeoff run. In 1932, *Fortune* magazine carried an article that listed the 10 worst autogyro accidents— none of them fatal.

So, what happened? Yes, the Great Depression and a world war were disruptive. But they disrupted many industries that nevertheless survived—notably the automotive industry itself. What happened to the autogyro as the basis for the universally owned flying car?

Gyroplanes didn't catch on as a segment of aviation for a number of reasons. The first and most obvious is that they were superseded by helicopters. A gyroplane is, in practice, a compromise between an airplane and a helicopter; one manufacturer advertises that it "does 90 percent of the mission at 10 percent of the cost." It's probably fairer to say two-thirds of the mission at one-third the cost. That sounds pretty good, but in today's specialized commercial market, the gyroplane loses to a market need for particular applications at the extremes: airplanes for speed and fuel efficiency; helicopters for hovering and vertical takeoff.

Still, the advantages of the gyroplane's compromises were, in Charles Zimmerman's words, "an airplane suited to the needs of the private owner." A private car is likewise a study in good-enough trade-offs.

## Roadable Airplanes

In the early days of the postwar world, it did indeed seem as if the Cessna Family Car of the Air would take off. Due to some federal pilot-training and training-aid programs, the number of licensed pilots in the U.S. increased from 50,000 in 1940 to 350,000 in 1950. It should be noted that both Germany and Japan had just been reduced to smoking ruins by saturation bombing, and the only existing defense against such an attack was to put a huge cloud of fighter planes in the air to shoot down the bombers. Wartime spirit still ran high among the public. In 1946, a record 35,000 general aviation planes were sold.[38]

This raises an important aside. Why can't we simply consider a private plane to be a flying car? The obvious answer is that it's not a car. Rather, it doesn't solve the "three vehicles problem." You still need a car to drive to an airport to fly your plane, and once you arrive at a new airport you need yet another car to get to your final destination. Let us consider the feasibility of a full-fledged car replacement: something you get into in your own garage or driveway and get out of at your final destination. The autogyro wasn't the only contender; in fact, in the '30s, '40s, and '50s there were numerous attempts at developing a car that could convert into an airplane.

The most famous attempt at a convertible car/airplane in the early 1930s was Buckminster Fuller's Dymaxion car. Its elongated egg shape was much more aerodynamic than cars of the day; it had a three-wheel stance, steered by its single rear wheel, as was standard for airplanes then. Fifty years later, Fuller explained the idea:

> When what we call a light plane, one flown by an individual, lands cross-wind, its fairing or streamlining makes it want to turn violently in the direction of the wind—the direction of least resistance. This is called ground looping. I realized that the most difficult conditions for my omni-medium jet-stilt superbly faired flying device would be when it was on the ground. What is popularly called the Dymaxion Car were the first three vehicles designed to test ground taxiing under transverse wind conditions.[39]

The Dymaxion did indeed inherit this problem from airplanes and was considerably less stable on the ground than ordinary cars. Furthermore, when operated as a car, it would tend to lose contact with the ground at high speeds, since its shape made it what today we would call a lifting body.

The first fully functional fixed-wing flying car was Waldo Waterman's 1937 Aerobile. Five years earlier, Waterman had built and flown the world's first tailless, or "flying wing," airplane. It was also the first plane with tricycle landing gear in the modern arrangement, with two wheels in the back and a steerable nosewheel. This is considerably more stable and resistant to ground looping in a crosswind than the old "tail dragger" configuration was.

The Aerobile was, by all accounts, easy to fly. It was also reasonably easy to drive. The wings could be removed and the fuselage driven around as a car. Five were built; unfortunately, on top of the difficulty of trying to sell them during the Great Depression, Waterman's financial supporter died unexpectedly, and Waterman had a serious illness for a year. And then, like almost every other major aviation figure, he was swept up into World War II.

It was the war that really clobbered the first age of flying cars. Despite even the Great Depression, the first great age of flying cars persevered during the '30s, and the technology advanced considerably. Without the disruption of global war, we might very well have produced a landscape of aviation considerably different from the one we got. Private aviation essentially vanished for the duration of the conflict. In addition to Waterman (who would end up teaching pilots), there were pilot and aeronautical engineers Moulton Taylor (who would produce the best-known flying car after the war) and Ted Hall (who had experimented with modular designs similar to Taylor's).

The Navy seized Pitcairn Aviation's airfield in 1942 under threat of eminent domain, paying Howard Pitcairn a third of its value. Pitcairn had developed the technology that made helicopters possible, and this was common knowledge in the aeronautical engineering world; the key patent was referred to simply as "582."[40] But, in a patriotic gesture, Pitcairn substantially reduced the royalties charged to any of the several new helicopter companies—Sikorsky, Piasecki (later renamed Vertol), Bell—

**Figure 13:** Moulton Taylor's Aerocar (National Air and Space Museum).

that were building helicopters under government contract. Under the terms of his agreement, the waiver expired in 1946.

As that year came around, and with the prospects of having to pay full royalties, the helicopter companies told the government that the price of helicopters would go up—so the government essentially nationalized the patents. Pitcairn sued. The suits went all the way to the Supreme Court, and the dispute continued from 1948 to 1977, by which time Pitcairn himself had been dead 17 years. He was legally vindicated in the end, as the court found that the government had infringed 59 claims in 11 separate patents, including, most notably, 582. Pitcairn's estate was awarded damages of $32 million. But Harold Pitcairn's (and Juan de la Cierva's) dream of a safe, low, and slow aircraft for general private use had been crushed.

By 1949, Moulton Taylor, returned from the war and free to dedicate his genius to his own business, had designed and built the flying car for the postwar world. Taylor's Aerocar was a two-seater coupe with a 143-horsepower engine that made it quite sporty on the road. (See Figure 13.) Unlike the Aerobile, the Aerocar had wings and a tail that folded up into a trailer that the car could tow. They could be attached to the car in about 10 minutes, forming a small private airplane that, by all accounts, was quite well designed, stable, and easy to fly. Four of the first model Aerocar were built; one Aerocar II, an improved design, was built; and one of the first-generation models was rebuilt into a third design, as Aerocar III.

The Aerocar finally got federal aircraft certification in 1956—unfortunately, just as the postwar piloting boom turned into a bust. Together, all the models flew a total of 9,000 hours. The Museum of Flight, which now owns the Aerocar III, lists it as having a cruise speed of 135 miles per hour. This would have been extremely useful in a world where the interstate highways had not yet been built. Even more useful, however, was the fact that the Aerocar solved the three-vehicle problem, which plagues us to this day. But even as Taylor perfected his Aerocar, forces arrayed against him. The

**Figure 14:**   Ted Hall's ConvAirCar (San Diego Air and Space Museum).

government had more need of intercontinental ballistic missiles than swarms of pilots, airlines wanted passengers, and the Eisenhower Interstate Highway System grew to cover the land.

A variant on Taylor's Aerocar was an airplane that was specifically designed to latch onto a car and carry it, but which wasn't capable of flying by itself. (That is, it didn't have a separate cockpit or landing gear.) This design begat the ConvAirCar, by Theodore Hall, which was explored, built, and flown at Convair in the late '40s. (See Figure 14.) Hall himself had been building and flying earlier prototypes before the war, efforts that went on hiatus when he was swept up in the war effort.

Convair put extensive resources into the ConvAirCar. The car part was custom-designed by Henry Dreyfuss, student of Norman Bel Geddes, to match the airplane part. Because of that, the whole machine was fairly close to standard private-plane specifications: It weighed about 1,500 pounds, carried 1,000 more as payload and fuel, had a 190-horsepower aero motor, and did 125 miles per hour in the air. Its major compromise was in ground mode: The separate car motor was only 25 horsepower.

The ConvAirCar very nearly succeeded. It was built by Convair, a major aircraft company. They had an ingenious marketing plan: They would sell you the car part, and a chain of dealers at airports would rent you the airplane part. When you bought the car part, you had a working car, whether you ever flew or not.

Unfortunately, test pilot Reuben Snodgrass watched the wrong fuel gauge while in the air, the airplane part ran out of gas, and Snodgrass crash-landed the prototype. Although he walked away from the accident, the resulting news photos of the smashed-up carplane caused so much negative publicity that Convair pulled out of the project, and Hall wasn't able to find another backer.

Had Snodgrass watched the proper gauge, or had Convair stuck with Hall, they would have solved the three-vehicle problem, but still not the runway problem. To this day, the closest solution to that remains the gyroplane. But what if we looked past Trafalgar Square and did one better than the pigeons?

## Helicopters

One of the first depictions of a vertical takeoff and landing (VTOL) flying machine in science fiction was also one of the first depictions of a (more or less workable) heavier-than-air flying machine of any kind, namely the Albatross in Jules Verne's *Robur the Conqueror*, published in 1886. It was essentially a clipper ship, held aloft by the direct thrust of a large number of airscrews substituted for the sails.

Could we build an Albatross today, and would it work? The answer is yes, although we would certainly make it lighter and more aerodynamically shaped than Verne's imagined open-decked wooden ship. In fact, in the years since 1886, a remarkable number of aircraft have been flown that use downward-pointing propellers for lift, beginning with helicopters.

Although helicopters in various experimental forms (and the word "helicopter" itself) had been around from the late 19th century, the craft was not developed into a practical machine until about 1940. When it was, it used many of the innovations that had allowed the autogyro to work, notably flapping hinges, damped lead-lag hinges, and collective and cyclic pitch controls.

Sikorsky's first working helicopter had three tail rotors: a side-facing one to counteract engine torque on the main rotor and control yaw, and two outrigger ones to control pitch and roll, under a main rotor hub that was essentially like de la Cierva's 1929 one. Sikorsky, with prodding from his Army sponsor, adopted Pitcairn's fully articulated hub using patent 582 and 10 other rotorcraft patents at Pitcairn's low wartime rates. This allowed him to dispense with the two up-and-down tail rotors, leaving only the sideways one in the standard helicopter configuration that we are used to today.

No runway. Point A to point B transportation in a single vehicle. So, why aren't we all in helicopters, one of the cornerstone ideas of the postwar world vision?

The main reason the helicopter didn't catch on to the expected extent is cost. One of the main components of the machine is the rotor hub, which even now is out near the edges of the envelope of technical capability. The hub is extremely complex. It also has to be extremely precise—tiny fractions of a degree or millimeters of position matter—and it is under tremendous stress. The blades pull outward with literally tens of tons of centrifugal force. Imagine trying to build a mechanical clock that had to be accurate to a second a day, and was small and light enough to carry in a suitcase, but which would have railroad cars attached as the hour and minute hands. So the rotor hub is not only expensive to manufacture, but, for safety reasons, must have constant maintenance by well-trained, intelligent, and motivated—read: expensive—technicians.

The obvious solution is to get rid of the rotor hub, using multiple small fans, like Verne's Albatross, instead of the one big one that requires the hub. The basic design of pointing fans at the ground can be enhanced by putting them in a shroud or duct. The

**Figure 15:** The Hiller Flying Platform.

simplest example of this is the Hiller Flying Platform, a sort of flying Segway which flew over 50 years ago.[41] (See Figure 15.)

The duct, which forms an aerodynamic structure known as a Kort nozzle, amplifies the lift of the propeller by some 40 percent. (It also cuts down on the noise.) The Flying Platform, like Verne's Albatross, used counter-rotating props. It was about seven feet wide, weighed about 370 pounds empty, and could lift another 180 with a total thrust of 555 pounds. It used a total of 80 horsepower, or one 40-horsepower engine for each propeller.

The version pictured was a five-foot prototype, which could not get out of ground effect, making it essentially a hovercraft instead of a flying machine. The seven-foot version, and an eight-foot version designated VZ-1 by the United States Army, could, but even so the craft had very limited altitude (33 feet) and speed (16 miles per hour).[42]

It turns out that the duct around the propellers is both a blessing and a curse. It amplifies the lift of the props significantly and provides stability, both very valuable features. The Flying Platform, especially the prototype five-foot version, was easier to balance than a bicycle. Helicopters, on the other hand, are inherently unstable; flying one is like balancing on a beach ball. With the duct around the fan, any tilt or sideslip induces forces that tend to put the platform straight again.

But, and this was the curse, that very stability limits its forward speed.

The Doak VZ-4 finessed the over-stability issue by tilting its ducted fans for forward flight. (See Figure 16.)

The VZ-4 worked quite well. As an experimental aircraft, of course, things were learned from the VZ-4's performance, and if it had gone into production it could have

**Figure 16:** The Doak VZ-4 (Courtesy U.S. Army Transportation Museum).

been optimized quite a bit. Even as is, however, it would have made a perfectly usable flying car. It weighed about a ton empty and a ton and a half loaded. It had a maximum speed of 230 miles per hour.

The Army went on to test several more related vehicles, notably the VZ-6 and VZ-8, or "flying jeeps," which were essentially two VZ-1's bolted together with the pilot sitting in between. It is worth noting that the Army, for whatever reason, decided to farm out the VZ-6 to Chrysler, which had never built any flying machine, and the VZ-8 to Piasecki, a rival helicopter company to Hiller. The VZ-6 was a complete flop (literally—it flipped over in the air, crashed, and was cancelled), but the Curtiss-Wright VZ-7, a non-ducted-fan quadrotor similar in configuration to today's toy drones, more or less worked. The VZ-8 was reasonably successful as a test vehicle. The Army Transportation Museum, where it resides in retirement, writes of it:

> The Airgeep II used twin 400-hp Turbomeca Artouste IIC turbo-shaft engines. . . . Despite the fact that the Airgeeps were intended to operate within a few feet of the ground, both were capable of flying at altitudes of several thousand feet. They were stable and able to hover or fly beneath trees or between buildings. In addition, the Airgeep was surprisingly effective as a weapons platform.

> Despite its many positive qualities, the Airgeep, like other ground effects machines developed during this period, was ultimately judged by the Army to be mechanically ill-suited to the rigors of field operations. The "flying jeep" concept was eventually abandoned in favor of further development of conventional battlefield helicopters.[43]

In the mid-1960s, about a decade later than the Army VZ program, the Navy experimented with a four-fan tilt-duct, the Bell X22-A. It was essentially a flying bus, being nearly 40 feet long and nine tons loaded. It proved once again that the tilting-duct concept was workable, being much easier to transition to forward flight than most of the other (remarkably numerous) experimental VTOLs of the 1950s and '60s. Like predecessors that had been built by the Army, the X22-A didn't quite make it as a weapons system; but it could take off and land vertically, and had a top speed of 255 miles per hour. In other words, a scaled-down version might make just the kind of flying car we're looking for.

## The Private Owner

In the 1930s, it was understood what would be needed in an aircraft "designed for the needs of the private owner." The autogyro, which could take off and land on a runway that you could easily put on a one-acre lot, might well have been just the thing to revolutionize private aviation. For one brief historical moment, we had the rare combination of the aeronautical genius, de la Cierva, and a leading businessman with substantial engineering experience and resources, Pitcairn, both dedicated to bringing the flying car to the people. Additionally, airplanes that would convert to a car and be driven to a destination from the nearest airport have not only been feasible but have been demonstrated, over and over, since the 1930s. If mass-produced, such vehicles could be within the means of a substantial fraction of families today.

But not in the '30s and '40s. Americans could not even afford cars during the Depression and the war. In the postwar world, the helicopter was the wonder of the age in much the same way that the autogyro had been in the '30s. Pitcairn had designed one helicopter during the war, the XR-9, which by all accounts was substantially more stable and easier to fly than the early offerings (e.g., Sikorsky's), as Pitcairn's engineering team had accumulated rotary-wing experience going back two decades. But Pitcairn stubbornly stayed with autogyros because they were substantially less expensive and were suited to the needs of the private owner. Then, instead of developing flying cars, he spent the remainder of his life and fortune in court, with the government fighting him tooth and nail.

By the 1970s, the mechanical and aerodynamic technology for flying cars of a variety of types—ranging from convertible airplanes to ducted-fan VTOLs—had reached the stage of successful experiments. These were mostly military and were, at the time, much too expensive for extensive private use. On the other hand, the same was true of computers. And like computers, there was no purely technological barrier to the continued development of flying machines. Another 50 years of improvement could very well have given us VTOL flying cars.

The bottom line is simple: The reason we don't have flying cars today isn't technological feasibility. We have had the means to build, manufacture, and improve flying cars for the better part of a century. If it's not feasibility, and if it's not technological know-how and innovative prowess, then what? That isn't the only question to keep you company the next time you lose a day to the three-vehicle problem. As you go from car to airport to airplane to airport to car, ask yourself: What else has slipped our grasp? What other life-changing possibilities could have filled the upper right half of the technological predictions chart?

> *Are we going to be stuck with the 55-mile-per-hour speed limit forevermore? Is this the fastest we will ever go?*

—**Moulton Taylor**

# 4

# Waldo and Magic, Inc.

In 1942, Robert A. Heinlein, under the pseudonym of Anson MacDonald, published the cover story in *Astounding*, the leading science fiction magazine. In it, he introduced the "Waldo F. Jones Synchronous Reduplicating Pantograph." Heinlein is recognized as the conceptual inventor of the telemanipulator, often termed a "waldo" for that reason. A waldo is basically what we would now call a robot arm, but instead of being programmed, it is operated via remote control by a human operator.

These came into use in the '50s, with the push provided by the rapidly developing science of cybernetics and control theory, and the pull stemming from the necessity of working with radioactive materials in the rapidly developing nuclear field. It is not as widely remembered that the original waldoes in the story were (a) self-replicating ("Reduplicating") and (b) scale-shifting ("Pantograph").

Waldo's goal in constructing a series of ever smaller waldoes is to be able to operate on individual nerve cells:

> Neither electromagnetic instruments nor neural surgery was refined enough to do accurate work on the levels he wished to investigate. But he had waldoes. The smallest waldoes he had used up to this time were approximately half an inch across their palms—with micro scanners to match, of course. They were much too gross for his purpose.... He used the waldoes to create tinier ones.... His final team of waldoes used for nerve and brain surgery varied in succeeding stages from mechanical hands nearly life size down to these fairy digits which could manipulate things much too small for the eye to see. They were mounted in bank to work in the same locus. Waldo controlled them all from the same primaries; he could switch from one size to another without removing his gauntlets. The same change in circuits which brought another size of waldoes under control automatically accomplished the change in sweep of scanning to increase or decrease the magnification so that Waldo always saw before him in his stereo receiver a "life-size" image of his other hands.[44]

Note that Heinlein even addresses, in an offhand science fiction kind of way, the problem of seeing what you are doing at smaller scales. But only in a science fiction kind of way; it's not a detailed plan of any kind. A better version would have to wait nearly two decades.

It seems almost certain that a young Richard Feynman would have read the short story "Waldo" or known about it. He had been, after all, an undergraduate at MIT, at the time perhaps one of the very biggest hotbeds of science fiction fandom. Princeton and Los Alamos, his subsequent intellectual homes, were not far behind.

Whatever his inspiration, in late 1959 Feynman, by then a professor of physics at Caltech who had already done the revolutionary work for which he would win the Nobel Prize, had been tapped to give the after-dinner speech at an American Physics Society conference. To have something fun but intriguing to catch the audience's interest, he dusted off the waldo idea and, being Feynman, went over it in detail and put a reasonably sound technological and scientific basis under it—and became impressed with the possibilities. In doing so he became the first to envision what today we would call nanotechnology.

This was so revolutionary that he told his audience, "In the year 2000, when they look back at this age, they will wonder why it was not until the year 1960 that anybody began seriously to move in this direction."[45] He missed it by a few decades; it took until the 1990s for people to begin to move seriously in that direction. By 2000, we had discovered a wide variety of electronic devices that worked at the atomic scale, including molecular transistors perfectly good enough to make a computer. Yet no nanocomputer or other complex circuit was built for a decade and a half.[46] We had the devices. What we did not have, and do not have still, is simply the infrastructure that macroscopic technology takes for granted: the ability to sort and test parts, to cut and join materials, to create frameworks that can hold devices in designed relationships, and to place parts into such frameworks. And yet we should have. In 1959, Feynman, in his now-famous talk "There's Plenty of Room at the Bottom," had described a straightforward, immediately actionable plan that, if it had been followed, would have resulted in exactly such an infrastructure well before 2000. Let us begin by quoting the relevant section of Feynman's talk:

> Now comes the interesting question: How do we make such a tiny mechanism? I leave that to you. However, let me suggest one weird possibility. You know, in the atomic energy plants they have materials and machines that they can't handle directly because they have become radioactive. To unscrew nuts and put on bolts and so on, they have a set of master and slave hands, so that by operating a set of levers here, you control the "hands" there, and can turn them this way and that so you can handle things quite nicely. . . .

> Now, I want to build much the same device—a master-slave system which operates electrically. But I want the slaves to be made especially carefully by modern large-scale machinists so that they are one-fourth the scale of the

"hands" that you ordinarily maneuver. So you have a scheme by which you can do things at one-quarter scale anyway—the little servo motors with little hands play with little nuts and bolts; they drill little holes; they are four times smaller. Aha! So I manufacture a quarter-size lathe; I manufacture quarter-size tools; and I make, at the one-quarter scale, still another set of hands again relatively one-quarter size! This is one-sixteenth size, from my point of view. And after I finish doing this I wire directly from my large-scale system, through transformers perhaps, to the one-sixteenth-size servo motors. Thus I can now manipulate the one-sixteenth size hands.

Well, you get the principle from there on. It is rather a difficult program, but it is a possibility.[47]

In a nutshell, Feynman's idea is to start from macroscale machining and fabrication and move to the nanoscale without ever losing the ability to do general fabrication and manipulation. That's exactly how Waldo did it in the Heinlein story. It is a possibility.

In working up the talk, Feynman got a glimmering of just how powerful a technology lay in that direction. He famously offered prizes for the first steps along such a pathway, $1,000 each for a tiny motor and tiny writing, hoping to kick-start interest. It may help to understand how important he considered it to realize that $2,000 in 1960 would have bought a brand-new Rambler American Deluxe two-door sedan. Both prizes were won, but, to Feynman's disappointment, nobody else really "got it," and interest petered out. Since then, there has not been a focused, coordinated effort to follow his pathway, or even a serious study of its feasibility.

## Nanotechnology

In 1976 or thereabouts, K. Eric Drexler, at the time yet another MIT undergraduate, had an inspiration:

I started thinking seriously about what you could build if you could design protein molecules and other biomolecules. I could see from the literature that there were all these mechanical and electronic widgets inside cells, that these things were synthesized chemically by the cells. . . . And I asked myself, Well, what if we could do things like that?[48]

1959, when Feynman gave his talk, was only six years after Watson and Crick had worked out the structure of DNA. Twenty years later, the notion of being able to understand, and possibly to use, the molecular machinery inside a cell was much better developed. By 1981, Drexler had worked out enough of an answer to his own question to publish a paper in the *Proceedings of the National Academy of the Sciences* on the subject, and to attract a cadre of other MIT students to form a Nanotechnology Study Group. By 1986, he and they had worked out enough of the implications to publish

*Engines of Creation*, the book which introduced nanotechnology to the wider world.[49] Drexler's and Feynman's respective approaches came from completely different directions: Feynman from the top down, starting with machines and getting to atoms; Drexler from the bottom up, starting with biochemistry and getting to machines. But as mature technologies they would get to the same point: molecular machines. Feynman spoke of the physicist being able to "synthesize absolutely anything;"[50] Drexler called it "complete control of the structure of matter."[51]

It's important to understand what they meant. A certain amount of common sense, and some perhaps not so common scientific sense, is necessary to interpret those phrases. After all, a stick of dynamite in the process of exploding is an arrangement of atoms, but don't expect to build such an arrangement one atom at a time. But such a too-literal interpretation of those phrases is no argument against the possibility of a general synthesis capability, reasonably construed.

So let's talk about reasonable things: a hamburger, for example. When you eat a hamburger, no atom is created or destroyed. They are merely rearranged. Some of the carbon atoms are now attached to oxygen atoms, which you exhale as $CO_2$. Others make up distasteful fluid and solid excreta—but, again, *they are exactly the same atoms as were in the hamburger.* Therein lies the possibility of a general synthesis capability. If you had the ability to rearrange atoms as you wished, supplying energy where necessary, you could transform the various waste products back into a hamburger.

In a sense, we do have a technology that does just that: We use the waste (including the $CO_2$) to fertilize a field of grain, some of which we keep and some we feed to a cow. We grind both the grain and the cow up into flour and hamburger, respectively, and cook them appropriately. We can imagine a machine that does the same thing directly, without using life-forms; but it must operate at the same molecular scales and do the same kinds of atomic manipulations that life does.

DNA is not magic. It cannot describe every possible arrangement of atoms. It can only describe the linear ordering of a string of constituent protein molecules (the amino acids). These, however, turn out to be able to form a vast array of molecular machinery, which performs all the basic functions of life. This is parallel to the limitations of our macroscopic technology. Don't ask a machinist to turn out the Mandelbrot set in high-tensile steel. But what he can turn out are all the gears, pulleys, shafts, and bearings needed to make another machine shop like his original one. And that's enough to form the technological basis of our modern world. Sometimes unstated but implicit from the beginning was the notion that nanotech would be a self-replicating technology. Heinlein said it explicitly—"reduplicating." Feynman followed suit, and Drexler started out with the self-reproducing mechanisms of life.

Actual cellular biochemistry has a number of well-known limitations. Cellular machinery—life—needs water to operate; it is limited to the temperature range between freezing and boiling. It operates slowly (compared to what's possible at the molecular scale) and is very low-power. Nanotech is not bound by these limitations; it could be much faster and higher-powered, and could operate across a much wider temperature range than can life. Properly appreciated, it holds out the potential of incomprehensibly tremendous capabilities. Consider that the difference between charcoal and diamond, between sand and computer chips, between diseased and healthy tissue, indeed between the dead and the living, is merely the arrangement of the atoms. No wonder Feynman found himself impressed. In *Nanofuture*, I described Drexler's vision:

> *Engines of Creation* is a technophile's dreamscape. It predicts microscopic replicating units able to build skyscraper sized objects to atomic precision. These could be buildings or they could be spaceships. It discusses artificial intelligence and engineering systems able to handle the enormous complexity such designs would require. It speaks of "easy and convenient" space travel, and describes a spacesuit so light and thin that you almost forget you're wearing it (present-day spacesuits are very awkward and arduous to wear and work in). It mentions cell repair machines and curing "a disease called aging." It talks about cryonics and resurrecting the frozen.[52]

Feynman had only talked about recording the *Britannica* on a pinhead, and maybe simplifying the process of complex chemical syntheses. Drexler had gone a long way further in working out what the ability to make "absolutely anything" actually implies.

Drexler, who knew he was onto a revolutionary idea, nevertheless kept his projections scrupulously inside the boundaries of conservative engineering estimates. It didn't matter, of course; once the concept of nanotech had gotten out, science fiction writers without the technical knowledge (or the scruples—after all, they were writing fiction) took off in all directions. By 1989, there was, for example, a *Star Trek* episode in which "nanites," very loosely based on Drexler's assemblers, which were, in turn, based on something like yeast, evolve into an intelligent species in a day or so—and then suddenly cast aside the slightest hint of being susceptible to evolution and act as a completely coordinated collective entity. Science fiction requires the suspension of disbelief, but that one required abysmal ignorance of 10 or so major areas of science and engineering at the same time. As enthusiasm for nanotechnology increased, so did the noise, and it became difficult for someone consuming the popular literature to understand just what was being claimed by those who were doing serious scientific research.

At the same time, though, the excitement inside the technical world was almost palpable. When I founded the Usenet nanotechnology discussion group SCI.NANOTECH in '89, it won approval by the largest vote (and largest margin of yes-to-no votes) that any Usenet group ever had. At the First Foresight Conference on Nanotechnology, the same year, more than one of the attendees remarked to me that it felt like being at one of the famous Solvay Conferences of the 1930s, which were so iconic in the development of modern physics.[53]

It is important to understand how much the thinking, and the excitement, about nanotech in the 1980s and '90s owed to Drexler as opposed to Feynman (or Heinlein). "Waldo" had been around for 50 years and "Plenty of Room" for 30—but no one had done anything about it. After *Engines*, interest took off exponentially.

# The Mightiest Machine

It is difficult, even for someone who has worked with these ideas for the past couple of decades, to get one's head around the utter raw power and potential of real nanotech. Consider the following, from Drexler's *Nanosystems*:

> The power density is large compared to that of macroscale motors: $>10^{15}$ W/m$^3$. For comparison, Earth intercepts ~$10^{17}$ watts of solar radiation. (Cooling constraints presumably preclude the steady-state operation of a cubic meter of these devices at this power density.)[54]

What Drexler is saying in this dry passage is that your thousand-horsepower ($10^6$W) flying car engine, for example, fits in a one-millimeter cube ($10^{-9}$m$^3$). It also means that one cubic foot of nanomotors would require more power (~$10^{13}$W) than all the machines used by the human race planetwide. Nanomotors to power a Kardashev Type I civilization (i.e., one that uses an amount of energy equal to all the sunlight that falls on the Earth, equal to $10^{17}$W), would fit in a 500-square-foot apartment (with eight-foot ceilings).

This is why *The Jetsons* is a more realistic depiction of the technology of 2062 than any of the "serious" futurism or science fiction. The reason that the world of *The Jetsons* is all buttons and levers and no gears or engines is because, in this kind of future, there simply won't be any visible motors. Motors will be everywhere, but they will be microscopic, like the motors in your muscle cells. Things will simply move, smoothly, silently, when you want them to, like your arms and legs do. And that is the less important part. The more important part is the effect of moving a self-replicating technology to the nanoscale.

All else being equal, the smaller a machine is, the faster it can run. A simple example is the nearly incredible miniaturization in scale and concomitant acceleration in speed that our electronics have undergone during the past half-century. Another familiar example is the wing-flapping speeds of birds and insects, from eagles to mosquitoes. The wings are moving through the air at a similar absolute speed, but since the mosquito has so much less distance to flap, it can do many more motions in a second.

In the real world, or, better put, the world of now, a part entering a typical factory may move at an average speed of five feet per second—walking speed—on its journey through the assembly process; and, in a typical factory, it has hundreds or even thousands of feet to go from loading dock to finished product. The rapidity with which a factory can process its inputs into outputs is clearly crucial to the value of the factory. A factory that can make 20,000 widgets a week is obviously worth two factories that can only make 10,000.

Imagine that an item of raw material needs to travel a mile through a factory, taking 20 minutes at walking speed. If the factory is scaled down to the size of an average cell, a couple of microns, the corresponding item gets through and into a finished product in a microsecond. In very simplistic terms, this is why humans reproduce in 20 years but *E. coli* reproduces in 20 minutes.

Reproduction speed has crucial implications for the overall economic growth rate. If your factory costs $100 million to build and can produce $3 million worth of widgets in a year, it will take you 33 years to regain the value of the factory. That level of productivity, 3 percent per year, has been an average for the Industrial Revolution over the past couple of centuries. (Note that if the widgets are not simply consumed, but reinvested, say, by improving the capabilities of the factory, the total value will double in 25 years instead of 33, due to the magic of compound interest.)

What, then, if the factory can make a copy of itself in a year, a month, a week? What about replicating itself in a day?

To give you something of a feeling for how vast the productive capability of nano-tech would be compared to current industrial technology, an anecdote. At a nanotech conference some years back, I was sitting with the nanotechnologist and writer Robert Freitas in the lobby of the conference hotel after the day's talks, chatting des-ultorily of this and that. At one point, the question arose of how long it would take to replace the entire capital stock of the United States, given a mature nanotechnology. By that we meant how long it would take to rebuild every single building, factory, high-way, railroad, bridge, airplane, train, automobile, truck, and ship.

At that point, Rob and I had each been thinking, researching, and publishing on the issues and techniques of molecular manufacturing for over a decade, and each of us had his favorite architectures, methods of analysis, and so forth. So we pulled out our notebooks and mulled and calculated for about five minutes. Then we looked up and answered each other almost in unison: "About a week."[55]

To someone who *hasn't* been studying the capabilities that can be reasonably pre-dicted for nanotech, that sounds fantastic, ludicrous, insane. It isn't. You are welcome to obtain a copy of *Nanosystems*, a high-end workstation, a pile of molecular-simulation software, an encyclopedia of mechanical engineering, and have at it. In 10 years or so of hard work, you will have a much better grasp of the subject.

Or you could just trust me with an analogy: The power and sophistication of our information-processing systems and devices today are demonstrably fantastic, ludi-crous, insane from the point of view of anyone in the '60s. The IBM 7094, costing $35 million in today's dollars, was used by NASA to control, coordinate, and navigate the Gemini and Apollo spacecraft from 1962 on. It had about 0.35 MIPS of process-ing power. In 2015, you could purchase for $35 a Raspberry Pi with 9,700 MIPS. The workstation I'm typing this on has 2,356,230 MIPS. *The same astronomical in-crease is possible with matter and energy, and we have known how to go about doing it since 1960.*

## From Bits to Atoms

Nanotech (as distinguished from "nanotechnology," which is essentially nanoscale surface and materials science) is a digital technology. Once we have machines built to atomically precise specifications—which can, in turn, build more machines to atomically precise specifications—we will have stepped onto the same escalator that we did in digital information technology.

My computer science mentor at Rutgers University had a quip: "To the first approximation, the hardware is free." What he meant is that the complexity of software required a lot of high-powered, and thus high-priced, brainpower. The complexity, and thus the software, was the bottleneck to increased capability. The complexity of software has grown like gangbusters, but it has, for the past 50 years, followed rather than led the capabilities, sizes, and speeds of the physical computers we run it on.

This will be true in spades with nanotech. In software, the substrate was the computers, running comfortably ahead of your programming prowess by virtue of Moore's Law. In nanotech, the substrate is physics. We have no lack of atoms. The "software" is the design of the machines, self-constructing in digital matter. You could physically rebuild America in a week, but how long would it have taken you to design it? And don't even think about getting construction permits!

Our physical technology is still in what we might call the Analog Age. The only field in which we are manipulating matter digitally is biotech, where nature has led the way. But within a decade or two the bridge will be crossed. We will begin to make machines that can make "absolutely anything," in the sense that a printer can print any page or a 3D printer can make any shape in plastic. But a few decades hence it will be shapes made from a wide range of engineering materials and with atomic precision. Among the things they will be able to make is more printers. At this point, the capabilities of nanotech will begin to take off on a Moore's Law-like curve.

Moore's Law is equivalent to a 60 percent growth curve. To match this in physical productivity, each machine need only produce a copy of itself, or, equivalently, do as much work as was involved in building it, in a year and a half. That is for the chip itself; once you factor in the rest of the computer and other constraints and bottlenecks, a good estimate for computer power growth over the past 50 years is a rate of about 40 percent.

A fairly conservative nanofactory design reproduces itself in an hour. The same kind of constraints and bottlenecks apply, but we might reasonably expect a 40 percent growth rate for pure physical manufacturing capability once full-fledged nanotech takes hold.

Compare the capabilities of your iPhone—and its price, size, and power consumption—to that of the IBM 1401, which filled a room, rented for $20,000 per month (in today's dollars), and came fully loaded with 16K bytes of hand-threaded ferrite-core memory. What's more, the iPhone comes with the equivalent of a movie camera, recording studio, record player, television, several radio stations, accelerometer, global positioning satellite downlink, telegraph office, and a library with thousands of books.

That will give you a basis for estimating what physical technology 50 years from now could look like compared to ours. Or, more to the point, it is what physical technology could have looked like right now, if we had listened to Feynman in 1959.

It is now decades past the year 2000, and, as we look back at that age, we wonder why it was not until long after the year 1960 that anybody began seriously to move in this direction.

A simple example will help us understand just how great a difference it might have made if we had. In early 2020, a novel coronavirus got loose and quickly spread around the world. In January of that year, Chinese scientists submitted the gene sequencing data of the virus for posting on Virological.org, a prepublication hub similar to the better-known arXiv.org. By February 24, Moderna, Inc., a biotechnology company pioneering messenger RNA (mRNA) vaccines, had used the sequence to create a vaccine and shipped 100 batches of the drug for early-stage testing.[56] In fact, the Moderna vaccine is the one that I got more than a year later. There are many compounded reasons why I had to wait a year, and they include some of the same reasons why 2020 was one of the most dismal years in recent history. One of them was simply bureaucratic inertia; we waited until December to receive permission to use the vaccines. But another, at least as important, was that we simply didn't have the physical manufacturing capability to produce them quickly in the necessary quantity. A SARS-CoV-2 mRNA vaccine dose is an arrangement of atoms. The Moderna one is given as two doses, each consisting of 100 micrograms of the active agent (as lipid nanoparticles) in half a milliliter of aqueous solution. Given that these are manufacturable in quantity with existing bulk technology, it is essentially a given that a nanofactory could make them with no trouble.

So here is your nanotech thought experiment: Suppose that, since 1960, we had made progress in nanofactories at the same rate as computers, so that anyone who has a computer now would have a nanofactory. Not a big thing, mind you; like a computer, it could be as small as a cellphone. Indeed, it would probably be a part of your smartphone. Just a quick update for your iDoc app, and it synthesizes your dose from the $CO_2$ in the air and a vitamin tablet.

*It is a possibility.* It has been a possibility for a long time. The mind-boggling elephant in the room is the very question Feynman asked: Why has it taken so long to move in that direction? If we had really grasped our opportunity, the entire physical paraphernalia of *The Jetsons'* world would be here *now*.

One possibility is that there is an Overton window effect in technology, a window into the world of ideas that frames what people are prepared to entertain, where ideas outside the window are not seriously considered. Really revolutionary ideas simply roll off men's minds like water off a duck. Machiavelli described the effect as "the incredulity of men, who do not readily believe in new things until they have had a long experience of them." Even such a respected and brilliant scientist as Feynman couldn't get people to take him seriously.

Still, as science fiction writer Edgar Pangborn put it, "I persist in wondering whether folly must always be our nemesis." Throwing away a technology that could have given us not only flying cars but immortality in peak physical health must surely count as folly. But perhaps there are other reasons for our failure.

# 5

# Cold Fusion?

# The Occurrence of the Impossible

*Petrified with astonishment, Richard Seaton stared after the copper steam-bath upon which he had been electrolyzing his solution of "X," the unknown metal. For as soon as he had removed the beaker the heavy bath had jumped endwise from under his hand as though it were alive. It had flown with terrific speed over the table, smashing apparatus and bottles of chemicals on its way, and was even now disappearing through the open window.*

—E. E. "Doc" Smith, *The Skylark of Space*

In February 1985, Kevin Ashley, a graduate student in chemistry at the University of Utah, found something of a commotion in room 1113 of the Henry Eyring Building, where the electrochemistry labs were. He describes the scene:

> The lab was a mess and there was particulate dust in the air. On their lab bench were the remnants of an experiment. The bench was one of those black-top benches that was made of very, very hard material. . . . The experiment was near the middle where there was nothing underneath. I was astonished that there was a hole through the thing. The hole was about a foot in diameter. Under the hole was a pretty-good-sized hole in the concrete floor. It may have been as much as four inches deep. . . .
>
> Stan and Martin had these looks on their faces as though they were the cat that had just swallowed the canary.[57]

Stan and Martin were, of course, Stanley Pons and Martin Fleischmann, and they were, on that day, extremely lucky men. They thought they were lucky because they had managed to obtain the fusion of deuterium into helium in an electrochemical beaker. They were actually lucky because if they had been correct, the neutron flux and gamma radiation from fusion producing that much energy would have killed ev-

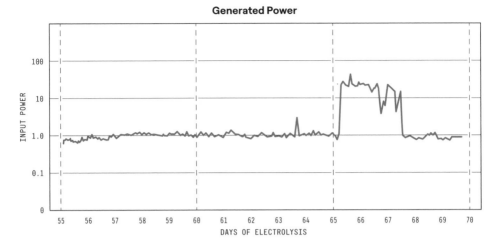

**Figure 17:** Trace from a Fleischmann and Pons cold fusion experiment from 1985–88: A calorimeter would run for two months producing the same power out as in, then for no apparent reason produce more than 10 times as much out for a few days, then just as inexplicably quit.[59]

ery living creature in the building and rendered it radioactive for years to come.[58] They were lucky that hadn't happened. But what had?

Fleischmann and Pons believed that they had managed fusion in a bottle, and proceeded to invest $100,000 of their own money—a quarter million in today's dollars—in further experimentation over the next five years. As well they might: If their experiments had produced a usable power source, they would have revolutionized not just physics and chemistry but the entire technological world. They would have had something that mainstream nuclear engineers could only dream of: a way to convert all the potential energy of deuterium to heat in a ridiculously simple and inexpensive apparatus. Or so they thought.

Four more years of experimentation left them out of money and without having much advanced their understanding of the phenomenon, whatever it was, that they had discovered. Over those four years, massive heat releases were rare. It was much more common for nothing at all to happen. When something did, the usual case was, after sometimes months of "loading" a palladium electrode with deuterium, it would produce a burst of heat, often 10 times more than could be accounted for by the electrolysis current, as measured by calorimetry. There would be no apparent reason for the burst to start, and none for it to stop, as it usually did within a few hours or days. (See Figure 17.) But what to make of these results? The two men were electrochemists—Fleischmann was arguably the second foremost physical electrochemist in the world at that point—and this was at the height of their professional expertise. It seems unlikely that they would have made major, systematic errors in measurement.

But in 1989 things changed. They needed funding and were trying to get it from the university. The university, of course, was keenly interested in possible patent rights to what might be the major source of energy for the next century. And they—the University of Utah administrators, as well as Pons and Fleischmann—were given extra reason to believe that they must be causing fusion by the work of Steve Jones, a nuclear physicist at nearby Brigham Young University. It is to Jones we owe the term "cold fusion."[60]

Jones used "cold fusion" to refer to the process of muon-catalyzed fusion (MCF), which is a demonstrated, expected result of perfectly standard physics. In MCF, a muon (which is a kind of big brother to an electron) and a deuterium nucleus form a pseudo-atom that is like deuterium—heavy hydrogen—but much smaller, so much so that another deuteron can get close enough to fuse with its nucleus. Muon-catalyzed cold fusion is completely accepted as real by mainstream nuclear physicists—but it doesn't produce as much energy as it takes to create the muons. So, as an energy source, it's a dead end.

The UU administrators were worried that they would lose priority for any patents arising from cold fusion, and perhaps Pons and Fleischmann themselves had visions of Nobel Prizes dancing in their heads. Whatever the reason, with Jones announcing his not especially surprising "cold fusion," Pons and Fleischmann were pressured against their better judgment into holding a press conference and rushing a paper into print. "I was not at all in favour of the high publicity route adopted by the University of Utah," Fleischmann later wrote, "and wanted to delay consideration of publication until September 1990."[61]

The ensuing press conference, and subsequent Preliminary Note published in the *Journal of Electroanalytical Chemistry*, entitled "Electrochemically Induced Nuclear Fusion of Deuterium," predictably became a media circus and stirred up a firestorm in the scientific world.[62]

## From Space Opera to Soap Opera

Your humble narrator was a young researcher at Rutgers University in those days, and I remember the excitement with a certain nostalgia. A friend at Bell Labs called to say—incorrectly, as it turned out—that the experiment had been replicated. A friend at Princeton had gotten a copy of the preprint of the paper, and as he was going out of town, left a much xeroxed and faxed copy for me at the desk of the Nassau Inn; I drove down and picked it up at midnight. The closest thing we had to a fusion expert in the computer science department was a numerical analyst who collaborated with some of the hot-fusion physicists at Princeton. I asked him what chance cold fusion had of working, and he quipped, "About as good as hot fusion!" Not many days later, I happened to be visiting the physics building, and there were a bunch of people, mostly grad students, trying to reproduce Fleischmann and Pons's results. They had set up an electrolytic cell, and they had a neutron detector, and they had surrounded the cell with a wall of lead bricks.

At that point, the incongruity of the whole business struck me. Back in Utah, Fleischmann and Pons, sure that they had fusion going on in their laboratory, were running their experiments on a lab bench in a plastic tub. And here in New Jersey were the physicists, skeptical that fusion was happening, running theirs behind heavy radiation shielding.

What happened next might have been predicted. In a sense, it was predicted by E. E. "Doc" Smith in his novel quoted above. Besides the bafflingly weird coincidence that both Seaton's fictional X and Fleischmann and Pons's palladium were rare platinum-group metals releasing energy via electrolysis, there was the sequence of subsequent events. When Smith's fictional scientist Seaton attempts to demonstrate his new discovery to his friends, it doesn't work and he is dismissed as crazy. Then he goes off to develop the idea on his own, and established industrial interests who want to maintain a monopoly on the generation of power send thieves to steal his work, hit men to assassinate him and his partner, and finally operatives to abduct his fiancée in a spaceship.

The novel's plot is only slightly more melodramatic than what unfolded in reality. Assassinations and abduction by spaceship stir the blood of young science fiction fans; in the real world, character assassination and the destruction of scientific reputation are less dramatic but more efficient, and better suited to the gravity of august learned bodies.

"It is in the admission of ignorance and the admission of uncertainty," Richard Feynman said, "that there is a hope for the continuous motion of human beings in some direction that doesn't get confined, permanently blocked, as it has so many times before in various periods in the history of man."[63] The practice of science, true science, science capable of seeing the unexpected and learning from it, must work this way. Bureaucracies with the word "science" in their names, not so much. In respectable scientific circles, including the editorial offices of most major journals, the patent office, funding agencies, and tenure committees of academia, cold fusion is leprosy-infested anathema. You can lose not just academic respectability but also funding, your job, and your career simply from being seen trying to keep an open mind on the subject. This sounds a lot more like "confined, permanently blocked" than it does an admission of ignorance and uncertainty by the scientific powers that be.

A review of the events of 1989 can put the physics community's coolness and incredulity toward cold fusion in context. In January of that year, two months before the March press conference in Utah, Congress had voted that the Superconducting Super Collider was to be built in Waxahachie, Texas. The SCSC had an ultimate price tag in the ballpark of $8 billion. The high-energy physicists were in a major political struggle to get the money.

In the midst of this contest, the discovery of cold fusion would have been a major embarrassment to high-energy physics. Given the zero-sum nature of science funding, there was a very real danger that congressional largesse toward big science might have gone to cold fusion research and not to a collider. But that wasn't all. To some extent, the high-energy physicists may have risked losing the cachet of being the smartest kid on the scientific block.

So, in April 1989, the Department of Energy convened the Energy Research Advisory Board (ERAB) and formed a committee to investigate cold fusion. The chairman and co-chairman were both from the world of nuclear weapons: John Huizenga was co-discoverer of two transuranic elements isolated in the detritus of the Bikini hydrogen bomb tests, and Norman Ramsey was the liaison from Los Alamos who oversaw the assembly of the atom bombs on Tinian island on their way to Hiroshima and Nagasaki. This is something of an indicator of the close ties between high-energy physics and high-power politics.

The ERAB committee rushed through a report that was widely regarded as dismissing cold fusion. But, on closer inspection, it seems clear that while Ramsey felt his job was to evaluate cold fusion, Huizenga felt his job was to discredit it. To that end, Huizenga wrote a popular book debunking it, which gained apparent authority because of his chairmanship of the committee. But the actual committee report is considerably more balanced. It stated, for example, "It is not possible at this time to state categorically that all the claims for cold fusion have been convincingly either proved or disproved."[64]

It appears that Ramsey had threatened to resign unless such language was included.[65] The report also lists 11 positive, along with 13 negative, experiments for excess heat. And the committee missed at least one credible positive result for excess heat, which occurred midyear at NASA Glenn Research Center but which was filed as a technical report instead of being published.[66]

One of the experiments that was listed, the one conducted at the Naval Air Warfare Center in China Lake, California, had not produced excess heat by the time of the report, and was listed as negative. But the researchers there did get positive results within a few months, and the principal investigator, Melvin Miles, is convinced to this day that the phenomenon is real.

(If you're counting, that comes to 13 positive and 12 negative results in 1989. But, as the ERAB report itself stated, "even a single short but valid cold fusion period would be revolutionary.")

So why was the reaction of the scientific funding apparatus so uniformly biased, given a fairly balanced review and scientists' moral duty of the admission of uncertainty? It was because once so much money, power, and prestige is involved, the game changes from true science to politics. And yes, this has happened "so many times before in various periods in the history of man."

## The Machiavelli Effect

Perhaps the most noted expression of this observation was by Niccolò Machiavelli in his 1532 masterpiece *The Prince*:

> It ought to be remembered that there is nothing more difficult to take in hand, more perilous to conduct, or more uncertain in its success, than to take the lead in the introduction of a new order of things. Because the innovator has for ene-

mies all those who have done well under the old conditions, and lukewarm de-
fenders in those who may do well under the new. This coolness arises partly
from fear of the opponents, who have the laws on their side, and partly from the
incredulity of men, who do not readily believe in new things until they have had
a long experience of them. Thus it happens that whenever those who are hostile
have the opportunity to attack they do it like partisans, whilst the others defend
lukewarmly, in such wise that the prince is endangered along with them.[67]

This happens so reliably, as Feynman and Machiavelli noted, that we should give it
a name. I call it the Machiavelli Effect. Perhaps the most reliable tell for the Machiavelli
Effect in science is when establishment scientists seek to prevent experiments that
would test a novel claim or hypothesis.

The Machiavelli Effect stood out more clearly the closer the critics were to hot-
fusion funding. For example, one of the most damning dismissals of Fleischmann and
Pons came from MIT's Plasma Science and Fusion Center—which tried a replication
of the experiment and reported no excess heat. But Peter Hagelstein, himself an MIT
professor, has pointed out that the MIT cells were very poorly designed, with power
measurement errors of 40 megawatts, whereas Fleischmann and Pons's had an error
of 0.1 megawatts. He wrote:

> For any unbiased scientist, it should be clear that the Fleischmann-Pons calo-
> rimetry was far superior to that reported by MIT. Therefore, the MIT calori-
> metric results cannot be used as a refutation of the Fleischmann-Pons reports
> of anomalous excess energy in Pd/D systems.[68]

If you look at the raw data from the MIT experiments, they actually do indicate an
excess heat release in the deuterium experiment, as compared to the light-hydrogen
control; but it was within their experimental error, so for publication the physicists felt
justified in adjusting the data to zero.

Given that literally billions of dollars of hot-fusion research funding were on the
line, the plasma physicists' failure to reproduce is less than convincing evidence that
Fleischmann and Pons hadn't discovered some real effect. It's even less convincing
given the emotional vituperation ("incompetent boobs," "cult of fervent halfwits")
heaped on Fleischmann and Pons by otherwise distinguished but elderly scientists in
the process of calling cold fusion impossible.

By November 1989, the ERAB report was out and the physics community was
firmly convinced that cold fusion was bunk. But it wasn't until November that
Fleischmann and Pons published the first paper that gave enough detail that a good-
faith replication effort could be attempted. By then, the Machiavelli Effect was on in
full force.

Hagelstein, mentioned above, gave a presentation in May 2012 about a cold fusion
experiment that had been running for six months with energy gains of up to 1,400
percent (albeit at very low power levels). His talk concluded:

> I recently had the experience of working with a large company in the U.S. who was interested in pursuing experiments in this area and helping out. So we . . . discussed with the technical people at this company of the possibility that they might put in some money. . . . So they got the agreement, they got the money, they got it to MIT, and we thought, "Good, now we can make some progress." However, a very famous physicist at MIT, who is involved in the energy program, found out what we were trying to do, and he cancelled the program. And he called up the vice president of the company and said some things that weren't very polite about the research. And not only did the funding not come and the experiments didn't happen, but my colleagues at the company were very worried about where they're going to work next.[69]

That Machiavelli predicted 500 years ago the actions of the leading lights of the physics community doesn't prove anything one way or another about cold fusion, of course. But it does mean that there should be less of a presumption that any partisan attacks were necessarily based on facts and sound scientific judgment. In Bayesian terms, one might say that there was a high prior probability of the attacks, independent of the soundness of the cold fusion reports.

One major problem with the state of cold fusion research today is that while there is a small cadre of smart and careful scientists slowly making small advances, there is also a substantial number of flakes, crackpots, and mountebanks making outrageous claims. Most of these don't make it into the research literature, but some do. Furthermore, due to the high difficulty of replication, it is all too easy for an honest but inexperienced researcher to fool himself into thinking he has attained positive results when he hasn't. This is an additional source of unreliability in the research literature. Careful studies have shown that only about 20 percent of published results could be replicated in, say, the biotechnology research literature.[70] I would be surprised if more than 10 percent of published cold fusion results were replicable.

Cold fusion is associated in the public mind—and in the minds of quite a few of its more vociferous fans—with perpetual motion, zero-point energy, antigravity, UFOs, and heaven only knows what other moonbeams and unicorns. Physicists, for some reason, have come in for more than their share of harassment from crackpots, ranging from flat-earthers to folks with their own versions of relativity or quantum mechanics. After seeing 99 of these crackpots—and they are endless—come in with their "limitless free energy" schemes (which spring from total ignorance of all the basic tenets of physics, such as conservation of mass/energy), it would require superhuman tolerance, not to mention patience, for a scientist not to dismiss the hundredth one that comes in making what seem to be the same claims.

But, of course, such patience is precisely what has occasioned most shifts into new paradigms. Arthur C. Clarke said, "The only way of discovering the limits of the possible is to venture a little way past them into the impossible." Virtually every technological and scientific aspect of the modern age that you currently enjoy is a consequence of someone having done that.

In 1989, Fleischmann was one of the top electrochemists in the world. If he and Pons had published a paper (and got a patent) detailing their apparatus, claiming minor and sporadic heat amplification (which is what they usually got) and not saying a word about fusion, they would have been much more resistant to attack from the physicists. Instead, the majority of their paper was about attempts to measure fusion products such as gammas and neutrons, an area in which they were not expert and which measurements they did incorrectly.

In doing so, they left themselves open to allowing high-energy physicists to have the say on the scientific validity of their discovery. Physicists attempting electrochemistry, themselves now working outside their own areas of expertise, largely failed to produce any excess heat. And they definitely didn't produce the kind of radiation that fusion should have.

At Utah University, support for cold fusion was high, and the National Cold Fusion Institute was founded. It ran through $4.5 million in funding, and, even with the direct involvement and support of Fleischmann and Pons, got no useful results.[71] The institute closed its doors within a year. For just that extra twist of the ironic knife, the state of Utah wound up paying out five times as much money to lawyers in the dustup over who was to blame as had been spent on the actual physical experiments.[72] Toyota funded Fleischmann and Pons to the tune of $40 million to continue work at a laboratory, IMRA Europe, over the next five or so years. Again they failed to produce anything useful. Similarly, Japan's Ministry of International Trade and Industry (MITI) spent $20 million on cold fusion research during the same period, also with no results.

Well, not *quite* no results. There was a slow, groping-in-the-dark kind of improvement in the ability to produce the effect, even though no one could do so reliably on demand. Fleischmann and Pons et al. completed five experiments at IMRA: experiment number three produced 150 percent excess heat, and number four produced 250 percent—and then number five produced zero.[73]

The experiments were incredibly finicky and maddeningly intermittent, and required enormous attention to detail—and then some luck. Richard Oriani, a leading chemist of hydrogen in metals, is quoted as saying that in his 50-year career, cold fusion experiments were the most difficult he ever performed. In other words, even if you have been enthused by my account here, don't dash out to your garage, grab some heavy water and palladium, and start bubbling—you're almost certain to fail. However, by 2008 there had been over 300 published replications and verifications of cold fusion phenomena by scientists around the world.[74] These are quite varied in scope and thoroughness, in the care and expertise with which they were done, and indeed in the credibility of the results. But, as the original ERAB report noted, only one of them has to be right for the effect to be real.

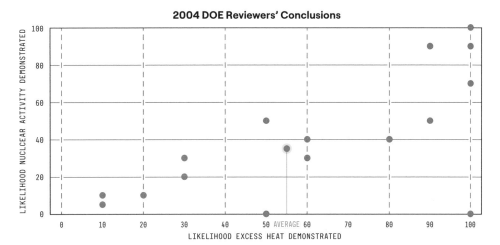

**Figure 18:** An estimate of reviewers' subjective probability (in percent) for cold fusion phenomena, as an interpretation of review texts. Two points overlay at 0,0. The point at approximately 55,35 is the overall average.

## Aftermath

Much more is now known about the conditions necessary for the phenomenon to occur. Few of the experiments performed in the fever of 1989 came close to loading the palladium sufficiently with deuterium. Later results by Michael McKubre (director of the Energy Research Center at SRI International) have shown that loading (that is, the ratio of deuterium atoms absorbed into the palladium to the palladium atoms themselves) has to be at least 90 percent, or nothing happens.[75] Fleischmann and Pons ran experiments for *months* before the palladium was properly loaded. Most of the early negative results had loading below 90 percent. Some pieces of palladium work, whereas other pieces, produced exactly the same way, don't. Palladium is so highly permeable to hydrogen that pieces of the solid metal are used as hydrogen filters. Fleischmann found that by far the most reliable palladium samples for the effect to work came from one manufacturer of hydrogen-filter palladium. Was it a stamping or rolling process that produced some unsuspected configuration of the crystal microstructure of the metal? Was it some parts-per-million impurity from the refining process? No one knows. Later researchers, such as Dennis Cravens and Edmund Storms, have found that only about 4 percent of samples, even from the manufacturing processes that do produce usable metal, actually work. And it takes a year of careful lab work to winnow out the four or so good pieces of palladium from an apparently identical set of 100. In 2004, the Department of Energy ERAB conducted another review of the status of cold fusion to determine whether to fund experiments in the area. The report was again noncommittal, neither strongly supporting nor denying the phenomenon. In an admittedly subjective exercise, I tried to estimate from each Department of Energy reviewer's remarks what probability he would have assigned to the two questions of whether the results to date had demonstrated excess heat or nuclear phenomena.[76] (See Figure 18.)

Again, while clearly not a resounding affirmation of low-energy nuclear reactions, or LENR, as cold-fusion-like phenomena are often referred to today, this is just as clearly not a dismissal either. The fact that the Department of Energy and the journals have given short shrift to new experiments seems to be more a politically driven policy than one reflecting the sober scientific assessment of their own report. (Note that NASA, the Navy, and the Defense Advanced Research Projects Agency, or DARPA, have continued to support research at low, sporadic levels.)

Suppose you make an unexpected scientific discovery—to pick an example, that a spider extruding a thread of silk as a parachute gives it an electric charge, which enhances its lifting power.[77] This is cool; it's neat; it's the sort of thing that scientists, real ones, live for. But you will not see a press circus and critical firestorm about it. The general public isn't interested. They yawn and say, "Nice, but so what?" Not so with cold fusion. The firestorm was ignited with prejudice. The difference, of course, is that the implications of the discovery appeared to be enormous. Part of this perception can be laid at the hands of the hot-fusion people, dating all the way back to AEC chairman Lewis Strauss and "power too cheap to meter." Fusion was going to be the energy source of the future, with fuel to be had from ordinary water. But with each failure, the capital cost of the next attempt kept climbing and climbing. (For example, the U.S. DOE has estimated the total cost of the current leading experiment, the International Thermonuclear Experimental Reactor in Europe, or ITER, at $65 billion.) Cold fusion promised a capital cost outrageously cheaper than any hitherto proposed fusion (or fission) technology ever dreamed of.

So, what ought we take from this soap opera of cold fusion and its discontents? It is not that cold fusion is a dead cert new energy panacea being suppressed by evil government and corporate men in black. I don't know any physics that makes it work. I feel much more comfortable saying that it is a solid hint that there is something in the physics *that we don't yet understand*. And I clearly see the repression of research, in a form I am all too personally familiar with, as we shall see.

The bottom line is that, because of the Machiavelli Effect, *we don't know* whether cold fusion is a possible source of energy. A cold shadow has been thrown into the heart of basic physics. Multiply that by the number of avenues of inquiry that might have led to substantive advancements and progress in the 21st century, but which are almost necessarily not aligned with the interests of the existing funding and science bureaucracy. Modern-day Simon Newcombs, who a century ago would have pontificated, "It can never work," have metamorphosed into bureaucrats who add, "And we'll ruin you if you try."

# 6

# The Machiavelli Effect

*The member of a mature scientific community is, like the typical character of Orwell's 1984, the victim of a history rewritten by the powers that be.*

—**Thomas Kuhn, *The Structure of Scientific Revolutions***

Like most people in the scientific world, by the end of 1989 I had accepted what had become the general consensus: that cold fusion had been a big mistake, based on misinterpretations and bad laboratory technique, and that there was nothing to it. It was only decades later, after having been personally introduced to the Machiavelli Effect in connection with nanotechnology, that I thought to reconsider.

In the case of nanotech, the systemic immune response to innovation was nowhere near as virulent as it was in the case of cold fusion. Instead of a cytokine storm, we got something closer to a sneezing hay fever. But for anyone close enough to the action, the essential aspects of the effect were plainly present.

Nanotechnology's star rose meteorically in the 1990s.

Drexler's 1991 MIT dissertation was the very first one explicitly on the subject of "nanotechnology." It was a formidably dense theoretical justification of a relatively small core of the manufacturing capabilities that nanotech should enable. When it was republished as *Nanosystems* in 1992 it became an improbable bestseller. The Foresight Institute's series of technical conferences on nanotechnology grew throughout the decade, attracting the top researchers in related fields. Prior to 1995, Google Scholar lists no references to Feynman's "Plenty of Room," which had been published in a Caltech magazine in 1960 and then in a book in 1961. In 1997, the Institute's keynote speaker was Richard Smalley, who had just won the Nobel Prize in Chemistry for his discovery of C60, the buckminsterfullerene molecule.

By 2000, the notion of "doing something" about nanotechnology had percolated up the political funding apparatus. President Bill Clinton, speaking at Caltech, proposed a National Nanotechnology Initiative. By then, the meme had latched strongly onto Feynman as a point of scientific respectability. Quoth Clinton:

> My budget supports a major new National Nanotechnology Initiative, worth $500 million. Caltech is no stranger to the idea of nanotechnology—the ability to manipulate matter at the atomic and molecular level. More than 40 years

ago, Caltech's own Richard Feynman asked, "What would happen if we could arrange the atoms one by one the way we want them?"[78]

And yet nobody—*nobody*—tried to do what Feynman had proposed. They didn't even try to do what Drexler had proposed. Instead, what happened was about as good an example of the pathologies of politically driven science funding as you can get. It created the classic setting for the Machiavelli Effect.

Remember that Machiavelli spoke of two separate classes of people. First, there are those who have done well under the old conditions, who have the laws on their side. These are the people who would, in fact, lose out or be discomfited by the innovation. Part of the dynamic is that this is also the group that tends to be politically and economically more powerful. This is a time-tested and inescapable feature of human society as we know it. For convenience, let us refer to this class as the Nobles, from the class in Renaissance Italy that Machiavelli observed.

The second class is the people who would have benefited from the innovation but who don't believe or understand the benefit would be worth the risk of being attacked "like partisans" by the Nobles. Again, this is quite normal, indeed quite rational, human behavior. Following our admittedly loose appropriation of Renaissance class designations, let us call this class the Tradesmen.

The NNI did not provide new money for a big push into nanotech. Rather, in the classic setup of a Machiavelli reaction, it shuffled money away from various established programs and researchers in related fields, mostly surface and materials science, who play the Nobles in the story.[79] The Tradesmen of the piece were the researchers who potentially might have worked on true nanotech, but who were working on other things or indeed still in school. They were, of course, completely unorganized. The Nobles reacted predictably: They labeled whatever they were already doing "nanotechnology."

They attacked Drexler's vision (and, by implication, Feynman's).

The results were predictable: From funding apparatus to researchers, everyone followed Machiavelli's script. The Nobles, already doing well under the old system, were threatened. The Tradesmen were hesitant to leap from the safe and familiar to the hard-to-believe new. To their incredulity was added the threat of partisan attacks. Innovation slowed before these combined headwinds. This dynamic is distressingly common—perhaps "endemic" would be a better word—in the zero-sum funding bureaucracy for science and technology.

It must be emphasized that the Machiavelli Effect has nothing to do with any conspiracy. After all, the existing researchers in this case were simply protecting their own turf. It really is more like the immune response of a social and economic system. What Machiavelli observed is an aspect of human nature, and the nature of human interaction. This is just an ordinary part of what Thomas Kuhn described as the operation of normal science. It has been observed by many, including Isaac Asimov:

I discovered, to my amazement, that all through history there had been resistance—and bitter, exaggerated, last-ditch resistance—to every significant technological change that had taken place on earth. Usually the resistance came from those groups who stood to lose influence, status, money as a result of the change. Although they never advanced this as their reason for resisting it. It was always the good of humanity that rested upon their hearts.[80]

That's almost as good a description of the Machiavelli Effect as Machiavelli made in the first place. But Asimov's "bitter, exaggerated, last-ditch resistance" doesn't quite capture the knife edge of Machiavelli's "attack like partisans."

So, by the early years of the 21st century, once the money had started moving, a classic Machiavelli Effect grew into being. The partisan attacks had the effect of endangering the funding of any researcher who attempted to work on anything vaguely like a molecular machine. There was, for example, a really nice and productive program studying nanotech—the real, atomically precise kind—at NASA Ames.[81] It was canceled. We at the Foresight Institute started having difficulty getting researchers who had presented interesting, productive work at earlier conferences to come back.

The intellectual content of the attacks was essentially nil. Smalley himself, who for whatever reason found himself in the vanguard of the partisan attackers, reversed himself more than once on the feasibility of self-replication at the molecular scale. First he agreed with it (in the days of his Foresight keynote), then he pooh-poohed it (with the implication that nanomechanical engineering was a foolish waste of time), and then he reversed field again, on the grounds that self-replicating nanobots would be too dangerous.[82, 83]

That sounds familiar. The fact that Machiavelli predicted 500 years ago the 20th-century actions of the leading lights of the nanoscale science community doesn't prove anything one way or another about nanotechnology, of course. But it bears repeating that we should not presume that every attack is based on facts and sound scientific judgment, and free of partisan influence. In Bayesian terms, the prior probability of the attacks was high, independent of the soundness of Drexler's (or Feynman's) vision.

Physical science and engineering are far from the only fields where the Machiavelli Effect can be found at work. Indeed, for various reasons they are probably the least, rather than the most, affected among the various fields of intellectual innovation. The great cognitive psychologist George Miller described how hard it was to break the stranglehold of behaviorism on psychology:

> The power, the honors, the authority, the textbooks, the money, everything in psychology was owned by the behavioristic school . . . those of us who wanted to be scientific psychologists couldn't really oppose it. You just wouldn't get a job.[84]

It is often difficult for an outsider to a field to distinguish between a case of the Machiavelli Effect and the more common case of the established experts pointing out that some would-be Galileo is, in fact, simply wrong. By far, the textbook knowledge in any given scientific field is, to steal a phrase from AI research, "probably approximately correct."

But in some cases—notably flying machines, but also the other ones with which Arthur C. Clarke fills the early chapters of *Profiles of the Future*—the consensus of the experts was wrong. The Astronomer Royal of Great Britain, for example, confidently declared rockets to the Moon impossible and the whole concept of space travel "utter bilge." The bureaucratic war on the mavericks was not only ill-founded but destructive of the overall search for scientific truth.

## Wright Flyer

In 1910, Wilbur and Orville offered their iconic invention, the first successful heavier-than-air flying machine, to the Smithsonian Institution.[85] They were turned down. They finally donated their Flyer to the Science Museum in London, where it stayed until 1948. In the interim, the Smithsonian exhibited Samuel Langley's Aerodrome and labeled it the first flying machine.

Langley had been director of the Smithsonian and had undertaken a major effort at a flying machine contemporaneously with the Wrights. The tests of the Aerodrome in 1903 had all ended in crashes, but in 1914 the museum lent the craft to Glenn Curtiss, who modified it and flew it successfully. Curtiss was a competitor of the Wrights and was keen to break their patent on the airplane.[86] Somewhat surprisingly, and somewhat tellingly, while the Wrights had spent about $1,000 of their own money to build the Flyer, Langley had received over $50,000 in government funding for his failures—the largest public expenditure by the U.S. for scientific research up until that time.[87]

Most everyone who is interested in flying is familiar with this story. Ironic feuds like this are the stuff of the all too human side of scientific history. But they are generally taken to be the exception, not the rule. Among scientists, it is an article of faith that basic research is a public good—that it provides much more overall benefit than it costs, but the benefit is spread so broadly across society that it is in no one's private interest to fund it. I fully believed this myself, and in fact spent the bulk of my career doing federally funded scientific research.

Research fads are funded and defunded all the time. In the 1970s, I had started out in artificial intelligence. By the '80s there was a wave of funding disinterest that researchers referred to as the "AI Winter." I had basically assumed that the people with the purse strings simply didn't understand what they were missing. After all, in the 1970s the government also funded research into space travel, nuclear power, even VTOL aircraft. A "miss" with AI would presumably be balanced by ongoing efforts elsewhere.

My time in nanotech convinced me otherwise. I don't think that the funding establishment ever turned on AI as viciously as the nanoscience establishment did on true nanotech. It took the complete, knowing, intentional betrayal of the basic concept of nanotech as envisioned by Heinlein, Feynman, and Drexler to make me question the competence of the top-down funding process to advance the interests of science, technology, and society. But, in retrospect, "free" money comes with as many strings attached as Gulliver on the shore of Lilliput.

The $50,000 in funding that Langley had gotten for the Aerodrome remained the exception rather than the rule in the United States up until 1940. Then the Manhattan Project spent billions of dollars, built more industrial plants than the entire U.S. automotive industry at the time—and succeeded. Defense-related research doesn't seem to affect economic growth much one way or the other per se, but the Manhattan Project did change hearts and minds. The remarkable success of the undertaking, and of the follow-on aircraft and guided-missile research, was influential in convincing people, including the futurists of the '60s, that federal research funding could accelerate scientific progress and economic growth.

Federal funding for defense research was entrenched by the time of President Dwight Eisenhower's 1961 farewell address warning of the military-industrial complex. The run-up in *non*-defense research spending, however, didn't happen until the 1960s, and during that decade the great bulk of it went to the space program.

And yet, as we have seen, the great innovations that made the major quality-of-life improvements came largely *before* 1960: refrigerators, freezers, vacuum cleaners, gas and electric stoves, and washing machines; indoor plumbing, detergent, and deodorants; electric lights; cars, trucks, and buses; tractors and combines; fertilizer; air travel and containerized freight; the vacuum tube and the transistor; the telegraph, telephone, phonograph, movies, radio, and television—and *they were all developed privately*.

It was not just the opinion of a few futurists such as Robert Prehoda, but the firm consensus of the entire economic and scientific establishment, that more federal money for scientific research could only help economic growth. Yet the evidence simply does not support the conclusion. England, where the Industrial Revolution began and which had experienced the most dramatic rise in individual well-being in history, had negligible public support for research throughout the 19th century, while France and Germany, which did have strong public scientific enterprises, never caught up. The United States did catch up, with our GDP per capita exceeding Britain's by the 20th century, but during that time we too were squarely in laissez-faire mode, perhaps even more so than Britain.[88]

If all the pundits are to be believed, at the very least, public funding of R&D shouldn't have *hurt*. And yet it coincided with the Great Stagnation, when observers agree that life-changing technological innovations slowed to a crawl.

A survey and analysis performed by the Organization for Economic Cooperation and Development in 2005 found, to the researchers' surprise, that although private R&D had a positive 0.26 correlation with economic growth, government-funded R&D had a negative 0.37 correlation!

The authors get fairly mealymouthed about it:

> The negative results for public R&D are surprising and deserve some qualifi-
> cation. Taken at face value they suggest publicly-performed R&D crowds out
> resources that could be alternatively used by the private sector, including pri-
> vate R&D. There is some evidence of this effect in studies that have looked in
> detail at the role of different forms of R&D and the interaction between them.
> However, there are avenues for more complex effects that regression analysis
> cannot identify.[89]

The report's authors and certain other economists in the field, such as Terence
Kealey, gravitate to the assumption that the phenomenon is a crowding-out effect,
with government funding crowding out private sector funding, as clearly happens
with sectors like schools. But my (and the other nanotech researchers') experience
showed that there is also a nontrivial backlash as well, one that I think clearly matches
the pattern of the Machiavelli Effect. I suppose you could call the Machiavelli Effect a
"complex effect that regression analysis cannot identify," but I think the explanation
is fairly straightforward. Centralized funding of an intellectual elite makes it easier
for cadres, cliques, and the politically skilled to gain control of a field, and by their nature they are resistant to new, outside, non-Ptolemaic ideas. The ivory tower has a
moat full of crocodiles.

## The Inventors of Tomorrow

> *Our Nation is at risk. Our once unchallenged preeminence in commerce, indus-
> try, science, and technological innovation is being overtaken by competitors
> throughout the world. . . . The educational foundations of our society are pres-
> ently being eroded by a rising tide of mediocrity that threatens our very future as
> a Nation and a people.*

—**National Commission on Excellence in Education, *A Nation at Risk***[90]

> *If it had been up to the NIH to cure polio, we'd have the best iron lungs in the
> world but we still wouldn't have the Salk vaccine.*

—**Samuel Broder, director of the National Cancer Institute**

Tyler Cowen, in *The Great Stagnation*, proposes that in the 1970s we ran out of the
"low-hanging fruit" of bright young people coming off the farm and into the university. It is true that there had previously been a long trend of leaving the farm, which
slowed in the '70s when virtually everyone had left agriculture.

Farm jobs in the United States as a percent of all jobs have plummeted, and as a
consequence there have been more and more young would-have-been farmhands
available to do anything else we need. Cowen's argument is that the supply of newly
freed-up farmhands dried up in the neighborhood of 1970, by which time the number

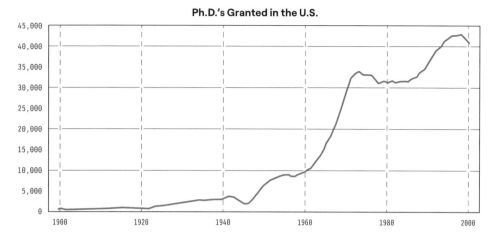

**Ph.D.'s Granted in the U.S.**

Figure 19:   Ph.D.'s granted in the U.S. (National Science Foundation).

of people working on farms had gotten so low as to be negligible. But the same thing was still happening with manufacturing jobs well into the 2000s. The U.S. has very consistently managed to produce more and more manufactured goods with fewer and fewer people, a trend that continues to this day—that means that we have freed-up factory workers available to do whatever else it is that we need.

We do not lack for workers. Indeed, the opposite is true.

If we took the same proportion of the workforce that was in manufacturing, agriculture, and the armed forces during World War II but that is today doing none of those things, and we committed them all to manufacturing with today's technological efficiency, we could produce five or six times as much as we do. In other words, we have huge headroom in what we could be producing (e.g., flying cars), if only we collectively happened to want to.

A further argument against "we ran out of talent" is that the feminist revolution happened at the same time as the Great Stagnation, roughly doubling the labor pool of smart, capable, young people. And women have, in fact, taken advantage of education: Of the 83,760 Ph.D.'s granted in the 1950s, nearly all went to men;[91] of the 440,497 Ph.D.'s awarded from 2000 through 2009, roughly half went to women.[92] Whatever the distribution, there were over five times as many Ph.D.'s per year as in the '50s, and 50 times as many as in the early part of the 20th century. Just how many smart young people do you need? Yet, over the same period, growth slumped and life quit changing for the better. Whatever was to blame, it wasn't a lack of smart young talent with college degrees.[93] (See Figure 19.)

The young people are there, and they are being educated in unprecedented numbers. If there is an argument to be made from education rates, it seems to run the other way—too many young people are spending too much time in the ivory tower, instead of doing real things in the real world. Kevin Jones, reviewing *The Great Stagnation* for *Mother Jones*, points out:

But there's a tension here that Tyler doesn't address. Technology grew like gangbusters in the first half of the 20th century, but it wasn't until the second half that education took off. So apparently it's not higher education that's really responsible for dramatic technological growth. But if that's the case, who cares about education? [94]

Given how well the huge bulge in Ph.D.'s in the 20th century seems to match the Great Stagnation, there seems to be enough evidence to adopt "ivory-tower syndrome" as at least a hypothesis. A proper investigation would require more time than we have here (and more economic expertise than your humble narrator can bring to bear), but it is simply stated: Too much education is bad for the economy, and inhibits technological innovation.

The philosopher Daniel Dennett, with a twinkle in his eye and a tongue in his cheek, points out:

> The juvenile sea squirt wanders through the sea searching for a suitable rock or hunk of coral to cling to and make its home for life. For this task, it has a rudimentary nervous system. When it finds its spot and takes root, it doesn't need its brain anymore, so it eats it! It's rather like getting tenure. [95]

In November 2005, while attending an Association for the Advancement of Artificial Intelligence symposium, I found myself in casual conversation with another AI researcher at a reception. One of the hot topics at the time was the Defense Advanced Research Projects Agency's Grand Challenge. The previous month, five autonomous vehicles—self-driving cars and trucks—had successfully completed a grueling 131.2-mile course in competition for a $2 million prize offered by DARPA, the defense department's R&D arm. This was a major advance, since the previous challenge, held just a year and a half earlier, had been a complete failure, with the best vehicle only managing to go 7.3 miles. [96]

I remarked as much to my AAAI friend, and he demurred. The apparent advance, he insisted, consisted of nothing but new ways of combining existing sensory, control, and navigation techniques. I elected not to pursue the matter, but his response sums up all too well the ivory-tower way of seeing things. If combining existing technology to accomplish the previously impossible isn't an innovative advance, what is?

Because that, of course, is exactly what the vast majority of actual technological progress consists of. However much a specialist may spy all the earlier efforts behind the parts and elements of a new machine, what the world notices is whether or not the machine works. The Grand Challenge results show graphically the difference between working and not working in the real world. In the case of self-driving cars, a major watershed was crossed between March 2004 and October 2005.

But there was no clever new trick that would have excited an academic—*nothing that even an active AI researcher attending a conference on AI recognized as an advance.* Academia is much more interested in "mind candy," intellectual tricks that impress other intellectuals with the intelligence of whoever thought them up, as contrasted with mundane techniques that just happen to work and do something useful.

## Failure of Foresight

In his classic work of mid-century futurism, *Profiles of the Future*, Arthur C. Clarke listed two major forms of failure risked by technological prognosticators. Clarke sets the tone for his book by spending two full chapters (and much of the introduction) discussing cases in which eminent scientists and pundits declared various technological achievements impossible, achievements that the 20th century then produced. He argued by implication that similar predicted limits currently fashionable might also be looked on with justified skepticism. He divided the mispredictions into two categories: the "Failure of Nerve" and the "Failure of the Imagination."

The Failure of Nerve applies when the facts are known: The science is there, the engineering understood, the pathway clear, and only the details remain to be worked out. But if the result is far enough out of common-sense experience, outside the box or the Overton window, the mind balks. The classic example of this, of course, was the near-unanimous insistence by the scientific community that heavier-than-air flying machines were impossible, a position that scientists continued to maintain for about five years *after* the Wright brothers' Flyer first flew in 1903. Simon Newcomb, in arguing that airplanes couldn't work because they would fall out of the sky when they stopped, committed Failure of Nerve, because he knew perfectly well one could travel fast on the ground, on a runway, at the beginning and the end of a flight. But, stuck inside the box, he assumed an aircraft had to act like a bird.

Arthur C. Clarke had plenty of experience with such spurious reasoning in the domain of rocketry. Perhaps the most flagrant case was the *New York Times* editorial denying that rockets could work in space:

> That Professor Goddard, with his "chair" in Clark College and the countenancing of the Smithsonian Institution, does not know the relation of action and reaction, and of the need to have something better than a vacuum against which to react—to say that would be absurd. Of course he only seems to lack the knowledge ladled out daily in high schools.[97]

Note the sneering arrogance of the editorial writer, and his complete and abysmal ignorance of what he was talking about. Now imagine that, instead of writing editorials, he was a grant manager at the NSF, with the power of academic life or death over 100 aspiring young physicists. How many of them would go on to produce the kind of innovations in rocketry that Goddard actually did?

With tongue firmly in cheek, the *Times* retracted the claim half a century later, as Apollo 11 was doing its midcourse correction on the way to the moon:

> Further investigation and experimentation have confirmed the findings of Isaac Newton in the 17th Century and it is now definitely established that a

rocket can function in a vacuum as well as in an atmosphere. The Times regrets the error.[98]

A modern example of the Failure of Nerve is nanotech: Absolutely standard physics, quantum mechanics, and chemistry predict that certain arrangements of atoms will behave as workable machine parts, and our existing knowledge in these disciplines allows us to calculate the performance of machines so built. The Failure of Nerve is essentially thinking inside too small a box. The walls don't really exist; you can walk right through them with your existing knowledge and techniques. Yes, it took the world-class genius of Feynman to break through the paradigm of what bulk technology could do, but the astounding capabilities he saw were of a technology completely accessible with the known laws of physics. By the time Drexler considered it, the false wall had been weakened by our understanding of the workings of the molecular machines inside the cell. But, to most, the box was still there.

Failure of the Imagination is, on the other hand, easier to understand and harder to prevent. It is essentially the same as Donald Rumsfeld's "unknown unknowns": where the thing that is making your prediction wrong simply isn't part of the tool kit of knowledge available to you. An exemplary case of this was Sir Ernest Rutherford's insistence that "the energy produced by the breaking down of the atom is a very poor kind of thing. Anyone who expects a source of power from transformation of these atoms is talking moonshine." Rutherford was the leading nuclear physicist of his day; indeed, he was the discoverer of the atomic nucleus. Yet he spoke before the discovery of the neutron and of nuclear fission, both unexpected within then existing physics.

Failures of the Imagination are much harder to spot, except in retrospect. All one can do is stack one's own imagination up against the experts', always a perilous proposition. But we can still reasonably call it a failure because we must always keep in mind that there are things we don't know. There is always an outside to the box, and the whole history of technology is of people finding ways to get there. Someone who played with a crystal radio set might well have allowed for the possibility of the transistor, not because he had any knowledge of quantum mechanics but by guessing that, given there was something in the physics of the solid state *that he didn't understand* that could act as a diode, there might well also be something that could act as a triode.

One of the great tragedies of the latter 20th century, and clearly one of the causes of the Great Stagnation, was the increasing centralization and bureaucratization of science and research funding. This meant that Failures of Nerve and Imagination, which are particularly prevalent among bureaucrats, went from being merely the incorrect predictions of pundits to causing resource starvation and the active suppression of progress. In short, such failures became self-fulfilling prophecies. A century ago, all the smart young people had to make their way in the real world, facing all the messy problems of life, work, and production, and some of them had imagination and nerve enough to invent new and better ways to solve them. Today, all too many of their descendants spend all too much time in the ivory tower, ever more dependent on handouts from the bureaucracy, and spend their time, efforts, and ingenuity inventing better ways to write grant applications.

This was evident enough by mid-century to be a worry sufficient for President Eisenhower to warn:

> The free university, historically the fountainhead of free ideas and scientific discovery, has experienced a revolution in the conduct of research. Partly because of the huge costs involved, a government contract becomes virtually a substitute for intellectual curiosity. The prospect of domination of the nation's scholars by Federal employment, project allocations, and the power of money is ever present and is gravely to be regarded.[99]

Measured by journal page counts, the amount of scientific knowledge we've accumulated in the past half-century has exceeded by more than a factor of 50 all the knowledge previously discovered. But it may well be that ivory-tower syndrome enhances apparent scientific discovery yet impedes useful application. The history of the Industrial Revolution offers us something of a natural experiment. As I summarized in *Nanofuture*:

> There was an amazing flowering of science in France at the dawn of the 19th century. D.S.L. Cardwell writes, "During the years 1790-1825 France had more scientists and technologists of first rank than any other nation ever had over a comparable period of time." We can mention Carnot, Lavoisier, Laplace, Montgolfier, Dulong, Petit, Biot, Fresnel, Gay-Lussac, Ampere, Savart, Fourier, Coriolis, Cauchy, and Lamarck—and these are just the ones whose names are attached to scientific laws and inventions that have survived to the present.

> What's more, the advancement of science and technology was a well-funded national policy. Sadi Carnot was a graduate of the École Polytechnique, an institution that had no parallel in England or anywhere else at the time. . . .

> So if you are a technological forecaster, what do you think happened next? What actually happened, of course, is that the Industrial Revolution occurred in Britain, not France. By 1850, Britain had railroads; Britain had steamships; Britain had the leading engine, machine tool, and textile industries in the world.[100]

The United States is the acknowledged leader in science and technology; our top research universities are not only among the best, but are the most numerous of any country. Our public policy dispenses substantial largesse for scientific research. But the effects of ivory-tower syndrome are on full display from the advent of the Great Stagnation to the present. The trees of knowledge are growing taller than ever, but someone appears to have been spraying paraquat on the low-hanging fruit.

# 7

# The Age
# of Aquarius

*The question had come into my mind abruptly: Were these creatures fools?*

—H. G. Wells, *The Time Machine*

In the 17th century, Isaac Newton discovered the laws of motion and gravitation, and invented differential calculus. He stood on the shoulders of giants, ranging from Aristotle and Ptolemy to Copernicus and Galileo, to Tycho and Kepler. In a winding but ultimately upward path, Western civilization had climbed from an understanding of the universe based on the lusts, envies, and rages of anthropomorphic gods to one based on logic and mathematics, observation and experiment.

It is difficult to express the beauty and value of science, or how rare and precious a thing it is in human history. Unlike other forms of knowledge, it builds on itself and gives value in our own lives to the hard-won insights of our ancestors. Ptolemy, with his epicycles, could tell you where the planets we had studied would be next year. But using Newton's celestial mechanics, Le Verrier could calculate from observations of Uranus where Neptune might be found. And there it was.

In unbroken succession, our knowledge and capabilities kept growing until the 1960s. Steered by celestial mechanics and powered by Newton's third law, which expresses the equal and opposite reaction, our rockets took us to the Moon.

But, at the same time, the heirs to that bequest of greatness were selling it for a mess of emotional pottage. The Age of Aquarius was dawning. Love, not calculus, would henceforth steer the stars.

In the 1974 movie *Flesh Gordon*, Emperor Wang of the planet Porno turns a "Sex Ray" upon the Earth, causing everyone to abandon whatever they may be doing, strip the clothing from anyone nearby, and indulge in torrid orgies. One is hard-pressed to distinguish the movie from the actual 1970s.

In a sense, H. G. Wells had predicted the phenomenon in 1895 in *The Time Machine*:

> It seemed to me that I had happened upon humanity on the wane. The ruddy sunset set me thinking of the sunset of mankind. For the first time I began to realize an odd consequence of the social effort in which we are at present engaged. And yet, come to think, it is a logical consequence enough. Strength is

the outcome of need; security sets a premium on feebleness. The work of ame-liorating the conditions of life—the true civilizing process that makes life more and more secure—had gone steadily on to a climax. And the harvest was what I saw!

What, unless biological science is a mass of errors, is the cause of human intel-ligence and vigour? Hardship and freedom: conditions under which the ac-tive, strong, and subtle survive and the weaker go to the wall; conditions that put a premium upon the loyal alliance of capable men, upon self-restraint, pa-tience, and decision. For countless years I judged there had been no danger of war or solitary violence, no danger from wild beasts, no wasting disease to re-quire strength of constitution, no need of toil. For such a life, what we should call the weak are as well equipped as the strong, are indeed no longer weak. Better equipped indeed they are, for the strong would be fretted by an energy for which there was no outlet. . . . This has ever been the fate of energy in secu-rity; it takes to art and to eroticism, and then come languor and decay.[101]

Wells foresaw not only the huge shift away from the age-old struggle to provide the bare necessities of life, but also the sexual revolution and the breakup of the fami-ly. The only thing that would have boggled his mind is that the transformation took just one century instead of the 8,000 he imagined.

## Art and Eroticism, Languor and Decay
## Make Love, Not War

It is generally accepted that Western culture—American culture in particular—went through a phase change in the 1960s and '70s. There were hippies and Wood-stock, environmentalism and Earth Day, and countless movements, including for free love, zero population growth, civil rights, pacifism, and feminism. Interestingly, many of the flower children of the latter-day bouquet echoed the sentiments and examples of the avant-garde Bloomsbury Set in turn-of-the-century London, perhaps without real-izing it. But Bloomsbury had been a tiny cluster of elites; in the 1960s, the transforma-tion was general, touching almost everyone.

The 1960s was a memetic Cambrian explosion. If we were to try to explain all the different social and cultural revolutions that flowered then in terms of their causes, we would face a bewildering multiplicity of coincidences. The various concerns of the flower children certainly did not appear suddenly. A century before the civil rights movement, we had fought a civil war over the question of slavery. There had been suf-fragettes since before the Seneca Falls Convention in 1848. The sexual revolution be-came general with the pill, but the free love movement dated back to at least the 19th century, and was championed by, among others, H. G. Wells himself and the Bloomsbury set. There had been conservationists since the founding of the national parks. There had been Marxists since Karl Marx and Luddites since Ned Ludd. The really interesting question is why all of these stirrings came to a boil in the '60s.

How could all of the revolutions have bloomed at the same time? When we see one flower in the desert, we ask about the plant. When we see them everywhere, we ask when it rained.

We can explain this to some extent using the hierarchy of needs that psychologist Abraham Maslow worked out as a guide to human motivations. At the bottom of the hierarchy are basic immediate needs, like physical safety, food, clothing, and shelter. Higher up are social needs, such as friendship and status; at the apex are things like spiritual fulfillment. Maslow's idea is that each of us is motivated to spend our time worrying and working toward the lowest level in the hierarchy that we haven't yet attained. The levels more basic than that (e.g., food) we tend to take for granted; the levels above our current state (e.g., spiritual enlightenment) are less important in comparison. Many have argued over which needs are in the hierarchy and in what order they are ranked, but the notion that there is some such hierarchy (perhaps differing in different people) has proved a robust hypothesis that remains in play seven decades after it was first suggested.

In the face of the culture shift of the 1960s and '70s, it seems reasonable to suppose that Western culture had succeeded in supplying the needs of the lower levels of the hierarchy, including the security of a well-run society. And with these levels attained, the modern Eloi could be thought of as all those Americans who became able to take certain things—the *Leave It to Beaver* suburban life sort of things—for granted. This means that they began to be able to spend more of their energy, effort, and concern on the love, esteem, consciousness-raising, and self-actualization levels.

It is important to emphasize, and reiterate, that this was not a bad thing but a good thing. The remarkable emancipation of women from the full-time drudgery of housework ("woman's work is never done"), which was accomplished by the labor-saving machines of the 20th century, isn't as celebrated as it should be—specifically because we now take it for granted! But by the '60s, housewives had been relieved of at least 75 percent of the work that they had had to do at the turn of the century. It was not a stretch at all to imagine the remaining 25 percent gone by 2062, producing Jane Jetson's button-pushing lifestyle.

There is a lot of truth to the stereotype seen in the inaugural *Jetsons* episode, in which Stella, the wife of George's boss, Cosmo Spacely, is off picketing the United Planets building with a sign saying "Martians, Go Home." Stella was, to some extent, a cultural stereotype dating back to the Bloomsbury Set. Even by 1962, she would have been the only one of the characters who actually led a life of complete ease with no real responsibilities. But that was beginning to change. Throughout the 19th and early 20th centuries, the innovative progress that began with the Industrial Revolution had taken hold, and productivity had risen so dramatically that even a lower-middle-class American in 1960 would have been considered substantially well-to-do a century before. Not only could we all be jet-setters; we could all be Stella—or Virginia Woolf.

## The Proud Tower

If you ever do much traveling in Europe, you are likely to visit preserved medieval cities such as Rothenburg, Germany. One of the features that tourists find so quaint about them is that they are surrounded by walls. But few tourists stand on the wall and imagine what it must have been like to defend it with not much more than an overgrown kitchen knife. Fewer still wonder where the attackers would have been coming from. When I asked one of the Rothenburg guides, I learned that quite often the marauders at the gate were just from the next town over.

The single most important fact about medieval life in Europe was the constant presence of war and the threat of it. The furnace in which Western civilization was forged combined a very strong evolutionary pressure for moral codes that optimized productivity, efficiency, and hard work, and ones that emphasized the needs of the polity over those of the individual. This is hardly an original observation: Both Hume and Montesquieu made it, in different words, in the 1740s. It's at least as hard to shape people as it is to shape steel. Hammer, tongs, and anvil are necessary. As George Orwell put it, "Sooner or later, a false belief bumps up against solid reality, usually on a battlefield." But by 1900, this blacksmithing had produced a manner of living that was obviously much better than it had been in previous centuries, and an understanding of solid reality, namely science, with fewer false beliefs than anyone had ever had before. Civilization, with a capital "C," was a thing, and it was a thing that specifically happened in Europe. We brought it to America largely prefabricated.

The hammer was put away in the postwar world. For all the angst about the Bomb and Mutual Assured Destruction (MAD), they drastically changed the nature of warfare. They short-circuited the evolutionary process. It was no longer the case that a society that slid into inefficient cultural or governmental practices was likely to be promptly conquered by the baron next door. The nuclear umbrella meant that the economic, political, and moral strength of a society was no longer at a premium. The processes of variation proceeded apace, but natural selection no longer operated, producing a Cambrian explosion of Eloi.

Beginning with Korea, the battles we fought were proxies done at arm's length; the United Nations forces could (and did) bomb all the industry in North Korea to rubble, but it made no difference, since the arms and materiel were coming from the USSR and China. But the nuclear umbrella prevented direct confrontation between the principals, and the back pressure on American society was considerably abated compared to World War II.

Peter Turchin, the statistical historian whose indicator trends all turned south in the 1970s, puts it this way:

> At any rate, there is a pattern that we see recurring throughout history, when a successful empire expands its borders so far that it becomes the biggest kid on the block. When survival is no longer at stake, selfish elites and other special interest groups capture the political agenda. The spirit that "we are all in the same boat" disappears and is replaced by a "winner take all" mentality.[102]

In the United States, a major factor in the zeitgeist shift away from our seeing technological accomplishments as a way to a better future was the Vietnam War, which came to be seen as both cruel and useless. It wasn't a "real" war, one that unified people behind a sense of national urgency. Indeed, the distrust and disdain for things military that it engendered was contagious enough to substantially reverse the admiration and pride Americans had in our technological prowess.

It is difficult to overstate how much the American experience in World War II shaped the culture and the institutions of not only the postwar world (and its expectations for flying cars) but the Great Stagnation and the graveyard of dreams. The individualistic and self-reliant culture of the 19th century had been winding down as the frontier filled up, but it gave way with a bang as Americans arose "in their righteous might" to prosecute the war—under a completely centralized bureaucratic government structure. This was obvious not only in organizing the military but in planning civilian production on an unprecedented scale. We came into the postwar world with the belief that such a structure *worked*. After all, it had not only won the war but left the U.S. the preeminent industrial power in the world.

In retrospect, the booming 1950s in America seem to be more a part of a dynamic trajectory than a sustainable steady state. It took the cooperative "same boat" spirit that the "greatest generation" inherited from the war to make the centralized corporate structures work. The baby boomers—my generation—split into two cultures, which, as far as I can see, not only didn't agree on values but fundamentally couldn't even understand each other. Ask any boomer what was the greatest, most pivotal event of 1969. Half of us will say the Apollo 11 Moon landing. The other half will say Woodstock. Both sets, hearing the other's opinion, will emit an honestly uncomprehending *"Huh?!?!"*

From the 1950s to the '70s, the average American followed the life cycle of Sinclair Lewis's *Babbitt*. In our millions, we went from conformity and cooperation to nonconformist rebellion, all in a search of personal meaning. The corporate state worked with the cooperating, self-sacrificing greatest generation. It didn't work so well with Eloi of the Age of Aquarius.

This is a key point that the golden-age science fiction writers did get wrong, a misjudgment that also answers for most of their missed predictions about technologies. From Wells to E. E. "Doc" Smith, to Asimov and Heinlein and van Vogt, it was a central theme of science fiction that not only technology but social decision-making would improve, become more scientific, and continue to advance along the lines that civilization had advanced for the past 1,000 years. Postapocalyptic novels, such as Asimov's *Foundation*, imagined that a collapse on the order of the fall of the Roman

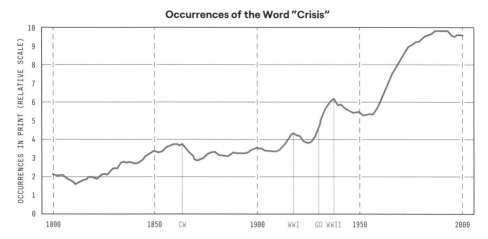

**Figure 20:** Eloi Agonistes in one graph: relative occurrences of "crisis" in print. Eloi angst about essentially nothing towers over the Civil War, World War I, the Great Depression, and World War II. (Data from Google N-grams.)

Empire would be followed by science perishing and superstition once again governing. But, in retrospect, it was ease and plenty and the lack of danger and struggle that engendered moral and intellectual decay.

War is no longer viable as a major shaper of culture and character. A global conflict fought with today's technology at the intensity of World War II would likely render the Earth uninhabitable. "Sooner or later a false belief bumps up against solid reality, usually on a battlefield," wrote Orwell.[103] What, then, is correcting our false beliefs? It is easy to look around today's public conversation and think, "Nothing."

## Eloi Agonistes

> *For though ours is a godless age, it is the very opposite of irreligious. The true believer is everywhere on the march, and both by converting and antagonizing he is shaping the world in his own image.*

—**Eric Hoffer,** *The True Believer*

If you ask the average person in the Western world today, they won't agree that they are the Eloi, living lives of perfect ease in a utopian garden. They will tell you that they struggle, that life is hard, that they are doing important, necessary jobs—saving the planet, even.

In Maslow's hierarchy, the next needs after safety are belonging and esteem. The pre-economic tribal environment from which the basic human psyche evolved was one in which esteem came from fulfilling a useful role and being appreciated for it, by others but also by oneself. This is why the human psyche is fairly resistant to the kind of life that Wells described the Eloi as living, a life as a sort of pet or lapdog that is provided for but does nothing useful for others. This is indeed a large part of the narrative power of the story (and its implied criticism of the "idle rich" of Wells's day).

Thus Wells only got the Eloi partly right. We know that he amended them to the Do-Nots upon a few decades' more experience. After all, the people who today take the good life of physical plenty and social stability for granted still need to find self-esteem, the support and respect of their peers. They need to believe that they are changing the world for the good, fighting the forces of evil, saving the planet.

It is the absolute genius of the postindustrial age that we have actually managed to meet that need. It is not that we have found our talented and motivated people jobs that actually make things better—the Great Stagnation story shows that, in fact, we are not doing nearly as good a job of this as our grandfathers did—but a huge segment of our current society *believes* that they are improving the world. We have managed to create a society in which a growing majority of people are doing useless, and in many cases counterproductive, work, but in which we all believe that we're doing something necessary and important, that we're making the world better. The planet has been saved more times than the heroine of *The Perils of Pauline*.

Nor have we subsided into the timid passivity of Wells's Eloi. For the lapdog lives of his fictional Eloi, Wells imagined that "the weak are as well equipped as the strong, are indeed no longer weak. Better equipped indeed they are, for the strong would be fretted by an energy for which there was no outlet." Perhaps in another 800 centuries we will have devolved down to that; but today our fretful energy presses hard at any possible outlet. In the Age of Aquarius, our focus largely shifted from conquering the physical world to conquering the interior worlds of other people.

In this postmodern struggle, objective truth is no longer an advantage. More often than not, truth proves inconvenient in the effort to make someone, including yourself, believe, using emotion instead of evidence.

It is widely unappreciated just how much we have evolved to deceive ourselves. By far, one of our most developed sensory and cognitive abilities is the ability to tell when we can trust someone: In a world of sometimes—but not always—trustworthy fellow beings, this is obviously a crucial skill, and one that would clearly be selected for in evolution. But it is also the basis for a classic evolutionary arms race: The ability to fool someone into thinking that you are being honest when you aren't would also be strongly selected for.

It turns out that, given the way the human mind is constructed, one of the best ways to convince someone you believe something is simply to believe it. (See, for example, the work of Rutgers University evolutionary psychologist Robert Trivers, and that of George Mason economist Robin Hanson.[104]) The contradictions between the thing that you have decided to believe and the facts that you might otherwise know can go ignored. Computationally, checking the consistency of all the facts in your mental database is expensive—why do it when it is more advantageous not to? One hardly has to rely on the efforts of these latter-day scientists, of course; it is one of the great themes of literature, ever since there was literature, that one of our greatest flaws as humans is our tendency to self-deception, hypocrisy, and hubris. From Don

Quixote to Elmer Gantry, the power to see ourselves as others see us is missing, but is often compensated for by an ability to get others to see us as we see ourselves.

T. S. Eliot was a longtime friend of Virginia Woolf, and was well acquainted with the Bloomsbury Group. But when she attempted to put together a "fellowship fund" to allow him to retire from his job as a bank clerk and write full-time, he rejected it, calling it "that cursed fund," and pointed out, "Half of the harm that is done in this world is due to people who want to feel important. They don't mean to do harm; they justify it because they are absorbed in the endless struggle to think well of themselves."[105]

We are built with an almost infinite capacity to believe things because holding these beliefs is advantageous for us, rather than because they are even remotely related to the truth. Half a millennium ago, we developed science not so much out of a love of truth, or even because we had begun to develop the tools to discover and propagate it, but because there had arisen a time and a place where it was more advantageous to know the physical facts of nature than not to. In other times and places, the relative advantages have been thought to shift.

The result of this is that we increasingly have major social institutions whose prestige comes in substantial part from virtue signaling rather than from actually producing useful results. In today's America, these institutions include, among others, health care, education, and environmental and safety regulation.[106] These are, unsurprisingly, the same spheres that are most afflicted by cost disease: The price that people seem to be willing to pay for them appears to be completely unconstrained by any consideration of value provided.

There is a psychological effect known as risk homeostasis, which posits that people have a particular level of acceptable perceived risk, and will act to keep this level roughly the same. The most common form of this is called risk compensation, or the Peltzman Effect, after Sam Peltzman, a University of Chicago economist who studied the phenomenon in the 1970s. His research found that as you introduced noticeable safety features, risky behavior shifted accordingly. For example, we introduced seatbelts in cars, and people drove faster and less cautiously enough to make the level of perceived risk stay about the same.

We can speculate that a more general form of risk homeostasis goes on in society as a whole. It seems clear that, during the 1950s, there was a holdover perception of exogenous military risk from World War II and Korea and the Cold War. But, by the end of the '60s, that had largely disappeared: People quit building backyard fallout shelters, schools quit holding air-raid drills, and so forth. But if people's risk-perception level is resistant to change, there are more ways than reckless driving to increase it. One obvious way is to start believing scare stories, from Chevy Corvairs to DDT to nuclear power to climate change. In other words, because the Aquarian Eloi were actually safer, they were more susceptible to any purported hazard, especially any one where being publicly worried would also fulfill the next Maslow level.

How well does this theory match the missed-predictions pattern that we saw with the Great Stagnation? Let's try a thought experiment: Consider a cellphone and a nuclear reactor. There will always be, for any technology, some entrepreneurial Eloi Agonistes attempting to make it appear dangerous and get it banned. In the case of cellphones, for example, they will tell you that the phone causes brain cancer by transmitting through your head. But, in the case of the phone, you have personal experience of it, you know many, many people who also have personal experience with it, and you see that there is no epidemic of brain tumors as the horror story would imply. But in the case of nuclear power, a technology with which very few of us have any personal experience, equally mendacious horror stories are widely believed. The further from everyday experience something is, the harder it becomes to see the truth instead of the horror story.[107]

You will remember that we found a remarkably strong correlation between energy intensity and the failure of predictions. The more powerful a technology, the easier it is to make it look scary, and the farther it is from the personal experience of most people.

## Only Man Is Vile

Bron Taylor, professor of religion and nature at the University of Florida, in his book *Dark Green Religion*, quoted John Burroughs, the great naturalist of a century ago: "If we do not go to church so much as did our fathers, we go to the woods much more, and are much more inclined to make a temple of them than they were."[108] He was right, of course: The influence of organized religion in social life and thought has waned considerably since his day, and an environmental movement that, at its extreme end, is well described by Taylor's title has replaced it. There appears to be a homeostasis in the human psyche for religion as well as risk.

In his book *Eaarth*, Bill McKibben wrote about walking along a creek in the Adirondacks and having it just feel wrong, not because there was any observable change from the natural state but simply because "merely knowing that we'd begun to alter the climate meant that the water flowing in that creek had a different, lesser meaning."[109] In one of the more widely quoted passages in his earlier book *The End of Nature*, he had described why that was bad: Instead of a world where rain had an independent and mysterious existence, the rain had become a subset of human activity. The rain bore a brand; it was a steer, not a deer.[110] It was bad simply because we had affected it; no further analysis necessary. No detectable change was necessary, only the fact that humans had had any effect on their surroundings was bad by definition. Indeed, it would be evil for us to even strive to *understand* it, for doing so would make it less "mysterious." This is the doctrine of original sin on steroids. There are plenty of Green fundamentalists who seem to think that the human race should simply commit suicide. For one example out of many, David Graber, reviewing McKibben, wrote:

> I, for one, cannot wish upon either my children or the rest of Earth's biota a tame planet, a human-managed planet, be it monstrous or—however unlikely—benign. McKibben is a biocentrist, and so am I. We are not interested in the utility of a particular species, or free-flowing river, or ecosystem, to mankind.

They have intrinsic value, more value—to me—than another human body, or a billion of them. . . . Until such time as Homo sapiens should decide to rejoin nature, some of us can only hope for the right virus to come along.[111]

This was written well before 2020. One wonders if he would care to rephrase it?

The Green religion has essentially superseded Christianity as the default religion of Western civilization, especially in academic circles. Since the 1960s, it has developed into an apocalyptic nature cult, centered around climate change. Green ideas have become inextricably intertwined with a perfectly reasonable desire to live in a clean, healthy environment and enjoy the natural world. The difference is, of course, that to most of us taking a walk in the woods, the human enjoying the natural world is a good thing, but to the fundamentalist Green he and all his works are bad. And if he is flying over the woods in a powerful machine, he must be downright evil.

The vast majority of the Eloi Agonistes who "believe in climate change" are not scientists who know anything about the enormously complex Earth climate system. They believe what they do because they have been told it by authority figures; they attach a moral significance to their beliefs; they believe that they are better people because they believe it. Eloi Agonistes react to skepticism not with careful self-searching reconsideration but with the scathing inquisitions of heretic-hunters. This is a recipe for religious faith, not scientific knowledge.

Because the issue of climate change is so politically polarized and people are so quick to categorize any statement on the subject on political grounds alone, it is incumbent upon an author to be very careful and specific about his position on the subject. My position is this: There is a truly enormous difference between Green activism, which is in its essence religious both in its moral basis and its methods, and the actual scientific study of the Earth's climate and its effect on the human race (and vice versa).

The religious side can be summed up by a 2017 poll that reported 39 percent of Americans believed there was a greater than 50 percent chance that climate change would spell the end of humanity.[112] This is an apocalyptic nature cult, indeed. But it is mind-bogglingly different from the actual scientific consensus on the subject.

The Intergovernmental Panel on Climate Change (IPCC) is the United Nations body responsible for collecting and summarizing the known scientific and economic data relating to climate and its effects on the economy. They are the organization that won the Nobel Prize, along with Al Gore, in 2007. They issue periodic Assessment Reports, which are, for all practical purposes, the official statement of the scientific consensus on climate change and its economic impacts. The lead-off paragraph of the economic-effects section in IPCC AR5, the current report as I write, sums up the issue:

> For most economic sectors, the impact of climate change will be small relative to the impacts of other drivers (medium evidence, high agreement). Changes in population, age, income, technology, relative prices, lifestyle, regulation, governance, and many other aspects of socioeconomic development will have

an impact on the supply and demand of economic goods and services that is large relative to the impact of climate change.[113]

*The impact of climate change will be small*—this is the scientific consensus.
*The existential threat of our time*—this is the religious catechism.

A vast gulf is fixed between these two sides. Climate change is a hangnail, not a hangman. In numerical terms, "the likely combined direct economic effects [of climate change] could reach 0.7 to 2.4 percent of the U.S. gross domestic product per year by the end of this century."[114] Anything, including energy conservation, that reduces the growth rate by a single percentage point will harm our grandchildren 50 times as much as climate change. The activists and media have created a hysteria that has absolutely nothing in common with the actual scientific knowledge on the subject.[115]

How can this be? How can such a huge gap in understanding persist, especially among those who consider it such an important subject? I think that it is an example of a well-understood phenomenon in public choice theory. This is often called Baptists and Bootleggers, from studies of the political support for Prohibition laws in the Southern states. How could two such apparently ideologically opposed groups both support the same laws? Of course, the Baptists supported Prohibition from true religious motives. But the Bootleggers had practical reasons: The laws formed a strong barrier to market entry and kept the price of their product artificially high.

So it is with the true believers of the Dark Green Religion and sane climate science. But there is an interesting flip in the dynamic. In the old South, all Bootleggers who agitated for stronger Prohibition must necessarily pretend to be Baptists and have religious motives. In the strange world of climate change, the opposite is true.

If the Dark Green true believers were to march under their own banner, holding as ineffable truth that for man simply to affect nature, even undetectably, is the essence of evil incarnate, their following would be limited. It would be like the Southern Bootleggers arguing for Prohibition because it would make them more money.

So the subterfuge runs the other way. This time, the Baptists pretend to be Bootleggers. The Bootleggers in this case are the scientists whose funding and prestige come from studying climate change, and the entrepreneurs who have businesses (e.g., renewable energy) whose revenues are enhanced by environmental concern, mandates, and subsidies. Thus we have the scientists saying that it's likely to cost us 3 percent of GDP, 100 years down the road, while vast numbers of Dark Green Eloi Agonistes pretend to read the science but claim that the very survival of the human race is at stake.

As it turns out, the two leading human causes of habitat destruction are agriculture and highways—the latter not so much by the land they take up, but by fragmenting ecosystems.[116] Given this, one would think that scientific Greens would be particularly keen for nuclear power; the most efficient, concentrated, high-tech factory farms;[117] and for . . . flying cars. But, in my experience, as was absolutely the case in the old South, there are plenty of Schrödinger's cats out there who are Baptists on Sunday and Bootleggers on weekdays.

## Ergophobia

> *Learn from me, if not by my precepts, at least by my example, how dangerous is the acquirement of knowledge, and how much happier that man is who believes his native town to be the world, than he who aspires to become greater than his nature will allow.*

> —**Mary Shelley,** *Frankenstein*

"Ergophobia" technically means a neurotic fear of doing work. As with other phobias, it implies that the level of fear is way out of line with any actual risk. But the term comes from the same Greek root as "energy," and you will find that physicists use "work" and "energy" to mean the same thing. So I feel justified in using the word to refer to the almost inexplicable belief that there is something wrong with using energy per se.

It is important to note that the origins of ergophobia in current American culture have little to do with climate change; the crusade against nuclear power preceded the rise of climate concern by two decades. There was even a period in the '70s when the activists were telling us that our cars and industry were emitting so much smoke that they would shade the Earth and precipitate an ice age. The climate movement today— the activist part, that is—looks very much more like a repackaged crusade against energy, using whatever excuses it can find, than an honest reaction to a newly appreciated enhanced greenhouse effect.

One of the central mysteries of the Green faith is the simultaneous belief that the Earth's climate is heading for a catastrophe of existential proportions, due entirely to human $CO_2$ emissions, and that completely emission-free nuclear power must be avoided at all costs. But why commit to such a contradiction? Even in a religion one must choose central mysteries with care.

The answer lies in recent history. It was the anti-nuclear-weapons movement that morphed into the anti-nuclear-power movement. Activists pushed the notion, and many probably believed it themselves, that power plants were giant nuclear bombs waiting to explode at a moment's notice and drench the world in fallout. As we have seen, the fact that this was completely untrue at the time, and has since become even less statistically plausible as greater safety protocols have been introduced, didn't seem to faze the activists in the slightest or impede the growth or influence of their movement.

We have already noted that both environmental and nuclear angst make sense when your touchstone is opposition to energy in any form. But why oppose energy on principle? Up until the 1960s, everybody understood how valuable, indeed crucially necessary to human well-being, it is. It seems likely that the fundamentalist Greens started with the notion that anything human was bad, and ran with the implication that anything that was good for humans was also bad. In particular, anything that empowered ordinary people in their multitudes threatened the sanctity of an untouched Earth.

## Meme Plague

The Xhosa are a southeast African tribe whose economy and culture were traditionally based on cattle-herding. In the spring of 1856, Nongqawuse, a 15-year-old girl, heard the voices of her ancestors telling her that the Xhosa had to kill all their cattle and destroy their hoes, pots, and stores of grain.[118] Channeling the ancestors, Nongqawuse explained that once this had been done the very ground would burst forth with plenty, the dead would be resurrected, and the interloping Dutch-German Boers would be driven from their lands. Surprisingly enough, the beliefs found fertile ground among the Xhosa and spread like wildfire, within months receiving the imprimatur of their king.

Many members of the tribe slaughtered their cattle; by the end of 1857, over 400,000 had been killed. The Xhosa also refrained from planting for the 1856–57 growing season, and as a consequence there was no harvest. No cattle, no harvest. It is estimated that 40,000 Xhosans starved to death and that about an equal number fled the country in search of food. By the end of 1858, three-quarters of the Xhosa were gone.

The Xhosa meme plague is clearly something of a dramatic outlier in the history of religious fervor run amok. But it stands as a stark reminder that when social feedback, superstition, hopes and desires, and the suppression of doubt and skepticism ("faith") line up, the resulting movement can make an entire people believe and do horribly self-destructive things that are completely at odds with common sense. In an individual, we wouldn't hesitate to call this insanity.

In modern cosmopolitan societies, the outcome of a meme plague is more likely to be a flatline than a complete collapse. But then it is a flatline we seek to explain. Back in the 1960s, Eric Hoffer noted that mass movements in America tend to become rackets, cults, or corporations. He has been misquoted often enough to create a stand-alone quip: "Every great cause begins as a movement, becomes a business, and eventually degenerates into a racket." This is probably the most succinct statement you will ever hear about the Eloi-Machiavelli Effect dynamic. The religious fervor of the crusade is necessary to satisfy the higher levels of the Maslow hierarchy. The crusades of the Eloi Agonistes, *especially* the ones that did significant good in their early days, or who were aimed at some incontrovertibly good end such as preventing nuclear war, succeed in the memetic ecology. Thence it is but a short step to the status of a widely enough believed religion to be a valuable support for the Nobles, and get the laws on its side.

One of the really towering intellectual achievements of the 20th century, ranking with relativity, quantum mechanics, the molecular biology of life, and computing and information theory, was understanding the origins of morality in evolutionary game theory. The salient point is that the evolutionary pressures on what we consider moral behavior arise *only in non-zero-sum interactions*. In a dynamic, growing society, people can interact cooperatively and both come out ahead. In a static, no-growth society, pressures toward morality and cooperation vanish: You can only improve your situation by taking from someone else. The zero-sum society is a recipe for evil; it exalts takers while suppressing makers. And the Green religion has as a central tenet achieving a zero-sum society.

Just as important is the corrosive effect that this position has on science. Zero-sum political funding of research has a strong tendency to bias scientists, creating an Eloi Agonistes view rather than a truth-centered one, and to amplify the Machiavelli Effect. Science is not the normal human way of thinking and operating; religion is. It has taken a 500-year battle, since Galileo, against the religious, authority-based, value-laden way of thinking to create a mental tool that gives us a chance to find actual, objective truth. But at the higher fervent-crusade levels of the Maslow hierarchy, the fervency subverts the honest, skeptical, value-free, single-minded search for truth that science requires. In one generation, we have come perilously close to destroying objective science and leaving love to steer the stars. Carl Sagan saw the same in his final book, *The Demon-Haunted World*:

> Science is more than a body of knowledge; it is a way of thinking. I have a foreboding of an America in my children's or grandchildren's time—when the United States is a service and information economy; when nearly all the key manufacturing industries have slipped away to other countries; when awesome technological powers are in the hands of a very few, and no one representing the public interest can even grasp the issues; when the people have lost the ability to set their own agendas or knowledgeably question those in authority; when, clutching our crystals and nervously consulting our horoscopes, our critical faculties in decline, unable to distinguish between what feels good and what's true, we slide, almost without noticing, back into superstition and darkness.[119]

Many of the classic science fiction writers depict civilizations that rise and fall, as real ones have done throughout history. Wells's Eloi have forgotten science. So has humanity in John Campbell's *Twilight*. Asimov's *Foundation* and Piper's Terro-Human histories are based on historical cycles at the scale of the fall of the Roman Empire. Heinlein's Future History series has something of a finer focus. It's much easier to predict that a galactic empire will decline and fall at some imaginary date than it is to pick a decade that will be marked by cultural tumult—even only 20 years ahead. Heinlein's Future History chart, first published in 1941, listed:

[Crazy Years] Considerable technical advance during this period, accompanied by a gradual deterioration of mores, orientation and social institutions, terminating in mass psychoses in the sixth decade.[120]

It's not hard to predict a deterioration of mores. Virtually everybody, as they get older, sees that happening, because mores change and we all judge the new ones against the ones we grew up with. Aristotle decried declining morals among the youth of his day—and Socrates had been executed half a century before for contributing to it. But Heinlein's prediction still stands out. What's more, he predicted a hiatus in space travel after what he calls a "False Dawn." The hiatus lasts for most of the 21st century, and is characterized by the "New Crusade," a religious dictatorship that suppresses scientific research and technological advancement.

Unfortunately, this is coming closer and closer to being one of the science fiction predictions that the writers did get right.

# 8

# Forbidden Fruit

Whatever the proximate causes of the cultural shift in the 1960s and '70s, the one incontrovertible fact is that it happened. By the '70s, otherwise sane men were wearing polyester leisure suits with plaid pants and wide polka-dot ties. The culture of trust and "same-boat spirit" of the '30s, '40s, and '50s evaporated in the culture wars of the '60s and '70s. This appears to have accelerated into the 21st century, but the latter-day consequences of cultural chaos are well outside our investigative purview, and, in any event, they happened too late to contribute to the reason we don't have flying cars.

What did contribute to the shift we can sum up succinctly: susceptibility to baseless horror stories and hostility to technology, and particularly to energy. Ergophobia grew from a rare neurosis in the '50s to a full-blown religion in the '70s. If this had happened to a small duchy in Renaissance Europe, the community would effectively have committed suicide like the Xhosa. Even in a free-market economy, as the U.S. had been in the 1800s, the scattered places where ergophobia took root would likely have gone bust, as so many of the religiously based utopian experiments did.

In Greek mythology, Cerberus is the three-headed dog who guards the gates of the underworld to prevent the dead from returning to the world of the living. The Cerberus blocking us from attaining the future we were promised has three heads as well. They are: the bureaucratic structure of science and technology, which amplifies the Machiavelli Effect; the ergophobic religion of the Eloi Agonistes; and the strangling red tape of regulation.

In latter days, these three forces have reinforced each other with dire synergistic effect; but historically, regulation all by itself was enough to doom early attempts at flying cars.

Let us take just one piece of technology that we know was possible, Moulton Taylor's Aerocar, and imagine what might have happened had it "taken off." Assume that the Aerocar, and competing designs such as Ted Hall's ConvAirCar, had simply followed the same path as the ground car did from 1900. To enhance plausibility, we'll start in 1950, at the height of postwar pilot-registration levels. For the first decade,

Aerocars would have remained relatively few in number, and—just as was true of cars—would have been owned mostly by rich dilettantes and early-adopter hobbyists. But by 1975, 20 percent of adult Americans (i.e., a majority of families) could have owned one. We can plausibly imagine that various improvements would have been made along the way: Perhaps we'd have developed a quicker and easier way of detaching and reattaching the wings, just as Charles F. Kettering's 1912 self-starter removed one of the more arduous and dangerous aspects of using an automobile. Higher power and higher speeds are almost a given. And, in a market economy, we can assume that a substantial amount of the improvement would have gone to increased reliability and ease of use.

By 1975, there could have been a whole new industry, some nontrivial fraction of the size of the automotive one. Not only would this industry have added to the growth of the GDP directly, but its indirect effects would have been enormous.

One way to gauge the impact that a flying-car sector could have had on the economy is to look at the impact that the Interstate Highway System did have. It's estimated to have been responsible for something like a third of all economic growth in the booming, optimistic 1950s. In 1996, on its 40th birthday, the government estimated that it had produced a trillion dollars in lower product prices, another trillion in time savings and operator costs, and $0.4 trillion in reduced fatalities, injuries, and accidents.[121] And that doesn't measure harder-to-quantify benefits. In 1986, the section of Interstate 78 across central New Jersey, which leads to New York City via the Holland Tunnel, was finally completed after being held up for decades in the courts. At the time, I owned a house in rural western New Jersey, about 15 miles from the highway. I could now get to New York in about one hour, whereas it had taken two before. The market value of my house essentially doubled. The value of my car, to me but not the market, also increased in ways hard to measure.

That's the kind of increase in the value of everything that we could have expected from flying cars. People would save time, enabling them to do more things. A whole new class of businesses, flying-car landing strips, would spring up, and other businesses would cluster around them. And the radius of economic activity of each household would vastly increase.

The interstate system clearly contributed to the robust 4.1 percent growth rate that the United States enjoyed up to the beginning of the Great Stagnation, at which point it began to run out of steam as a source of increased productivity. Flying cars would have been able to finesse that slowdown in two ways: First, their practical speed range is much greater than earthbound vehicles, so more places are brought closer together. Second, the highways hit a limit where the cost to build a main road to less populated places was more than the economic benefit generated. To connect an isolated small town directly to the high-speed mainstream with the Aerocar requires a much smaller investment, in the form of a landing strip.

Arguably, then, just this one invention—the Aerocar specifically but convertible aircraft in general—technology that has been built and proven to work, could by itself have ameliorated the Great Stagnation.[122] If we had taken that path, given the kind of advances and innovations that have happened in other areas of technology, we would almost certainly have something at least halfway "Jetsonian" by now.

But we didn't; and the question is why.

## A Multitude of Officers

"He has erected a multitude of New Offices, and sent hither swarms of Officers to harass our people, and eat out their substance."[123] So declared Thomas Jefferson and the Continental Congress in 1776. Eight score and nine years later, Winston Churchill echoed the sentiment, decrying "vast bureaucracies of civil servants, no longer servants and no longer civil." The problem is perennial.

In May 1908, Wilbur Wright arrived in France to make public demonstrations of the Wright Flyer. When he opened the shipping crates containing the airplane, however, he was horrified to discover that the contents were a pile of junk. He wrote home with barely restrained umbrage, detailing the horrible packing job that had led to the machine being extensively damaged en route.

Luckily for the Wrights' domestic tranquility, it turned out that French customs officials had opened the crates for inspection on the docks at Le Havre, rummaged through the meticulously packed parts, casually tossed them back in, and sealed the crates with a few random nails. All the damage had taken place during the overland trip to Le Mans.

Wilbur, not to be stopped by that kind of misfortune, made friends with, and enlisted the help of, Léon Bollée, an aviation enthusiast who had an automobile factory. The two spent the summer essentially rebuilding the Flyer. The public demonstrations that secured the Wright brothers' worldwide reputation as the true inventors of the airplane took place in August.

In mid-century America, Bruce Hallock was a pilot, aeronautical engineer, and entrepreneur who had seen Waterman's Aerobile as a young boy and been captivated by its tailless design. He set out to build a flying car with the least number of parts to add and remove between conversions.

To finance his invention, which he called the Road-A-Plane, Hallock sold an airplane he owned. It was a Noorduyn Norseman, a single-engine bush plane made in Canada, which had been used for light cargo duties by the Army in World War II and then sold as war surplus. The buyers of Hallock's Norseman were a well-established missionary group that distributed the Wycliffe Bible throughout South America. To deliver the craft, Hallock flew the Norseman to Lima.

Sometime later, Hallock was rousted out of bed in the middle of the night by a federal marshal and dragged off to jail. The Norseman was deemed a "weapon," and Hallock was accused of violating the Neutrality Act. The jury quite reasonably found him not guilty. And though his Road-A-Plane ultimately flew successfully, Hallock, having spent quite a bit on lawyers, never found backers or attempted to manufacture it in quantity.[124]

**Figure 21:** Bruce Hallock and the Road Wing. (Courtesy Austin Hallock.)

The other nearly successful flying car of the postwar era was Robert Edison Fulton Jr.'s Airphibian, which was a car-plus-wings design like the ConvAirCar. The Airphibian also appears to have run into a regulatory swamp. First tested in 1946, by 1950 it had yielded five prototypes, which had logged 100,000 miles in the air and 6,000 car/plane conversions. Historian Andrew Glass describes it:

> In 1950, Fulton flew his prototype Airphibian to Washington, DC, and landed at National Airport. From there he drove directly to the headquarters of the Civil Aeronautics Administration to claim the first official certification to begin production of a flying car in the United States. . . . However, Fulton's financial backers had become discouraged with the seemingly endless expense of meeting government production standards, and they withdrew their support.[125]

When asked why his Aerocar had failed to "take off," Moulton Taylor had a ready answer. It wasn't the size of the nascent market, or his need of a partner with more business savvy or money. Taylor insisted that the problem was government regulation.

As late as 1975, he was in serious negotiations with Ford to have the Aerocar mass-produced. It was the Federal Aviation Administration and the Department of Transportation that threw a monkey wrench into the negotiations. Seven years earlier, Taylor had already secured an airworthiness certificate for the Aerocar. Airworthiness wasn't the issue. He claimed that the agencies turned thumbs down on the Aerocar "because everybody would have one, and we couldn't handle the [air] traffic."[126]

A clamoring public might have overcome the agencies' failure of nerve. But by then the wave of hostility toward and suspicion of technology that began in the 1960s was sweeping America.

The Motor Trend Car of the Year in 1960 was the Chevrolet Corvair. The Corvair, with its rear-mounted air-cooled engine, was a mild innovation, and even that only for American drivers. The most successful car model of all time, the VW Beetle—as well as the Porsche 911, hardly a slouch in the handling department—had the same configuration. Most owners, among them my uncle in the '60s and Jay Leno today, loved it. Studies by Texas A&M, the NHTSA, and others concluded that the Corvair had, if anything, slightly superior handling compared to other cars of its period.[127]

The Corvair was the first American car with independent rear suspension, something that most cars have today. It was introduced to compete with the VW Beetle, which had the same swing-axle configuration. But because it was new to American drivers, mechanics, and automotive engineers, it was an opportunity for classic Eloi Agonistes' entrepreneurial fearmongering. This particular mongering was spearheaded by Ralph Nader, who made his name and fortune with the book *Unsafe at Any Speed.*

In the 1920s, when the big uptake in automobiles was getting underway, most people had a family member or a friend on a farm or in a factory who was familiar with machinery. Having a knowledgeable person accessible meant that you wouldn't simply believe scare stories. But by the 1960s this was not nearly as often the case. More people were susceptible to scare stories because there were so many fewer ripstop threads in the fabric of society.

And a public that isn't skeptical of scare stories is a public susceptible to the fearmongering of muckrakers and the promised safety nets of regulators.

## The Great Explosion

> *If you want to start an automobile company in this country you need a handful of engineers, and at least 1,000 lawyers.*

> —**Arnold Kling**

The Great Stagnation coincided with a rise, nay, a flood, of federal regulation.

Administrative law scholars refer to the Nixon administration of the 1970s as the "regulatory explosion."[128] The '70s saw the passage of the Clean Air Act, the Clean Water Act, and the National Environmental Policy Act, as well as the creation of OSHA and the EPA, the Consumer Product Safety Commission, and many more. The number of pages in the *Federal Register* increased by 121 percent under Nixon. Nor did it stop there: According to a report commissioned by the Manufacturers Alliance for Productivity and Innovation (MAPI), the number of regulations for manufacturers went from about 1,500 in 1980 to 18,500 in 2011.[129]

Today, the federal regulatory code is over 175,000 pages long. It's not light, casual reading either. Here is Federal Air Regulations Part 91, section 1443, paragraph (c):

> (c) Notwithstanding paragraph (b)(3) of this section, after maintenance, preventive maintenance, or alterations performed by a repair station certificated under the provisions of part 145 of this chapter, the approval for return to service or log entry required by paragraph (a) of this section may be signed by a person authorized by that repair station.[130]

It's not about flying. It's not about aircraft. It's not about maintenance of aircraft. It's not about who is qualified to do maintenance. It's an addendum, with three separate references back into the regulatory corpus, to the rules regarding *who is allowed to do paperwork* about maintenance. The entire Federal Aviation Regulations/Aeronautical Information Manual, which every airman is responsible for knowing, is 1,150 pages long. At least it was in 2018; a new one comes out every year.

In 2013, economists John Dawson and John Seater published a study in the *Journal of Economic Growth*, "Federal Regulation and Aggregate Economic Growth."[131] They calculated that, if we had simply maintained the level of regulation that we had in 1949, the median household income in America would now be $185,000. Instead, it is $53,000.

This is only striking because we don't intuitively understand exponential growth. It represents nothing but the compounded effect of a 2 percent difference in overall economic growth rate over the intervening six decades. Dawson and Seater basically found that 2 percent differences in growth rates are readily observed between countries, and that these differences correlate robustly with regulatory regimes. While such studies often avoid consideration of the positive benefits the regulation might have had, it is hard to imagine that the average American family would readily pay $132,000 a year for them. And with cause. The actual benefits of regulation are almost always overestimated. A classic analysis of the 1962 drug laws, for example, found that they cut the rate of the introduction of new drugs in half while having no measurable effect on average quality.[132]

It is commonly believed that regulation of things like medicine and cars is necessary to ensure safety. But life expectancy in the U.S. rose from 47.3 to 68.2 in the first half of the 20th century, an increase of over 30 percent; however, in the latter half, the age of regulation, it rose only from 68.2 to 76.8, just 11 percent.

What's more, the regulatory regime has burdened aviation long and oppressively. Any element that goes into a certificated airplane costs 10 times what the thing would otherwise. (As a pilot and airplane owner, I have personal experience of this.) But perhaps the most obvious demonstration that regulation unnecessarily strangles an industry is the rare case where an area actually becomes deregulated. This happened to the commercial air carrier business in the U.S. by way of the Airline Deregulation Act of 1978, a response to the fact that in-state airlines operating solely in Texas and California had fares about half those for comparable flights in the rest of the country,

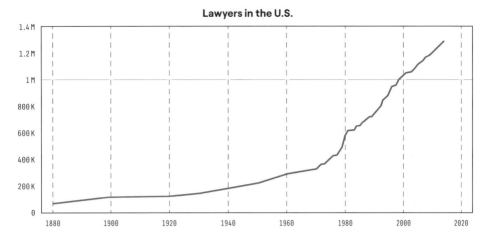

**Figure 22:** Lawyers in the U.S. (American Bar Association)

where interstate airlines operated under the tight regulation of the Civil Aeronautics Board. The result:

> The United States Airline Deregulation Act of 1978 was . . . the first thorough dismantling of a comprehensive system of government control since the Supreme Court declared the National Recovery Act unconstitutional in 1935. . . .
>
> Between 1976 and 1990 average yields per passenger mile—the average of the fares that passengers actually paid—declined 30 percent in real, inflation-adjusted terms. . . . The savings to travelers have been in the range of $5 billion to $10 billion per year.[133]

Air travel is unequivocally safer now than it was before deregulation. Moreover, by taking intercity travelers out of cars, the low airfares made possible by deregulation have saved many more lives than the total number lost annually in air crashes.

The 1970s brought an increase in product liability, a major social change that was nominally aimed at safety. Scare stories worked on juries just as well as on the reading public, and juries would vote enormous awards for accidents that didn't have any reasonable connection to malfeasance on the part of a manufacturer. The clarion call of so much money had a predictable result; the number of lawyers exploded in the 1970s. (See Figure 22.)

This led directly to the collapse of the general aviation industry. Over the course of the 1970s and '80s, it was strangled by the explosion of product-liability lawsuits. From a history of Cessna, the company whose name is most synonymous with private planes in the public mind:

> In 1985 Cessna was bought by General Dynamics Corporation and in 1986 production of piston-engine aircraft was discontinued. Over 35,000 172s had been manufactured up until that point.[134]

The company cited product liability as the cause for their demise. The corporation's CEO, Russ Meyer, said that production would resume if a more favorable product liability environment were to develop.[135] Piper, the Ford to Cessna's Chevy among private plane manufacturers, went bankrupt in 1991.

The *Concise Encyclopedia of Economics* explains why:

> Until the 1980s, property and liability insurance was a small cost of doing business. But the substantial expansion in what legally constitutes liability has greatly increased the cost of liability insurance for personal injuries. The plight of the U.S. private aircraft industry illustrates the extent of these liability costs. Although accident rates for general aviation and for small aircraft declined steadily, liability costs for the industry soared, so that by the 1990s the U.S. private aircraft industry had all but ceased production.[136]

As I noted before, there had been a dip in sales of private planes during the energy crisis, but it had bounced right back thereafter. This is good evidence that the later collapse was a systemic change and not a cyclic downturn. Even Congress managed to note that an entire industry had been destroyed, and in 1994 passed the General Aviation Revitalization Act, which limited product liability to planes that were younger than 18 years old.

This helped a tiny bit: There was a small bump in aircraft sales in the late 1990s. But the fact remains that a new Cessna Skyhawk cost $25,000 in 1980 and over $300,000 today.[137] And then the Great Recession of 2008 destroyed whatever was left of the industry; as I write, only 700 small private airplanes are shipped each year. That's just a single airplane for every half-million Americans.

General aviation was just one industry. A bit more visible and a bit more vulnerable than most, it was no exception. No useful and valuable, and thus profitable, industry or occupation escaped the feeding frenzy. Just ask a doctor how much he pays in malpractice liability insurance.[138] According to a study conducted by Tillinghast-Towers Perrin, the cost of the United States tort system consumes about 2 percent of GDP, on average. If we assume that this mostly started around 1980, when the number of lawyers skyrocketed and the airplane industry was destroyed, the long-run compound-interest effect on the economy as a whole is startling: Without it, our economy today would be twice the size that it actually is.[139] And that doesn't even include the consequences of taking more than a million of the country's most talented and motivated people and putting them to work making arguments and filing briefs instead of inventing, developing, and manufacturing.

If, in 1900, Orville and Wilbur Wright had faced the regulatory and legal environment that we have now, the Flyer would never have gotten off the ground. Indeed, we would never even have gotten the family car.

Wilbur Wright said, "If you are looking for perfect safety, you would do well to sit on a fence and watch the birds."

Benjamin Franklin might have added, "Those who would give up essential Liberty, to purchase a little temporary Safety, deserve neither Safety nor flying Cars."

## Miracles

Economist Robert Gordon described the great burst of productivity and innovation in the early 20th century as a miracle, with the implication that the Great Stagnation was simply a lack of miracles. But we know how to produce an economic miracle on demand. One happened on Sunday, June 20, 1948, in Germany. The online *Encyclopedia of Economics* describes it:

> After World War II the German economy lay in shambles. The war, along with Hitler's scorched-earth policy, had destroyed 20 percent of all housing. Food production per capita in 1947 was only 51 percent of its level in 1938, and the official food ration set by the occupying powers varied between 1,040 and 1,550 calories per day. Industrial output in 1947 was only one-third its 1938 level. Moreover, a large percentage of Germany's working-age men were dead. At the time, observers thought that West Germany would have to be the biggest client of the U.S. welfare state; yet, twenty years later its economy was envied by most of the world. And less than ten years after the war people already were talking about the German economic miracle.[140]

What had happened? Under the Allied occupation, Ludwig Erhard, a keen follower of the Austrian-British economist Friedrich Hayek, had come to hold a position which might be described as economic czar. In one fell swoop, he cut the Gordian knot of regulation that had been strangling the economy: rationing and wage and price controls, many of which had been originally imposed by Hitler and Göring.

The *Encyclopedia* continues:

> The effect on the West German economy was electric. [Henry] Wallich wrote: "The spirit of the country changed overnight. The gray, hungry, dead-looking figures wandering about the streets in their everlasting search for food came to life."

> Shops on Monday, June 21, were filled with goods as people realized that the money they sold them for would be worth much more than the old money. . . .

> In June 1948 the bizonal index of industrial production was at only 51 percent of its 1936 level; by December the index had risen to 78 percent. In other words, industrial production had increased by more than 50 percent.

> Output continued to grow by leaps and bounds after 1948. By 1958 industrial production was more than four times its annual rate for the six months in 1948 preceding currency reform. Industrial production per capita was more

than three times as high. East Germany's communist economy, by contrast, stagnated.

Over the long run, unchecked regulation destroys the learning curve, prevents innovation, protects and preserves inefficiency, and makes progress run backward. All the reasonable predictions of the golden age of science fiction and postwar techno-optimism were predicated on the continued growth of energy and continued sanity in law and regulation. Both presumptions held throughout the Industrial Revolution up to the 1970s. And then they didn't.

Robert Heinlein put it like this:

> Throughout history, poverty is the normal condition of man. Advances which permit this norm to be exceeded—here and there, now and then—are the work of an extremely small minority, frequently despised, often condemned, and almost always opposed by all right-thinking people. Whenever this tiny minority is kept from creating, or (as sometimes happens) is driven out of a society, the people then slip back into abject poverty.
>
> This is known as "bad luck."[141]

In an illuminating side note to the German economic miracle story, General Lucius Clay, the U.S. commander, is supposed to have called Erhard up after the announcements and said, "Herr Dr. Erhard, my advisers tell me you're making a terrible mistake."[142]

And Erhard replied, "General, my advisers tell me the same thing."

All that was needed for a miracle was for the right-thinking people to be ignored.

Looked at under the hood, the Great Stagnation was really the Great Strangulation. There is a difference. A stagnation might elicit a resigned shrug, a murmured "Boy, it would have been nice to live during the years of low-hanging fruit, before the stagnation set in." A strangulation calls for Sherlock Holmes and Hercule Poirot.

The stunning progress of the Industrial Revolution from the 1700s to the 1970s wasn't a miracle but the result of a large enough percent of enough generations being determined to be greater than the last and innovating their way, through technology and science, toward being greater in fact. They were able to do so in part because the entangling brambles of regulation that snag us at every turn largely weren't there. Conversely, the Great Stagnation that followed didn't happen to us; we did it to ourselves.

The really interesting question is whether we can undo it.

## The Family Car

In 1902, H. G. Wells penned a book, remarkably prophetic in many respects, that can be taken as the definitive fin de siècle take on the probable course of the 20th

century, at least as far as technology is concerned. It was called *Anticipations of the Reaction of Mechanical and Scientific Progress Upon Human Life and Thought.*
Wells wrote:

> There will, first of all, be the motor truck for heavy traffic. Already such trucks are in evidence distributing goods and parcels of various sorts. And sooner or later, no doubt, the numerous advantages of such an arrangement will lead to the organization of large carrier companies, using such motor trucks to carry goods in bulk or parcels on the high roads....

> In the next place, and parallel with the motor truck, there will develop the hired or privately owned motor carriage. This, for all except the longest journeys, will add a fine sense of personal independence to all the small conveniences of first-class railway travel. It will be capable of a day's journey of three hundred miles or more, long before the developments to be presently foreshadowed arrive. One will change nothing—unless it is the driver—from stage to stage. One will be free to dine where one chooses, hurry when one chooses, travel asleep or awake, stop and pick flowers, turn over in bed of a morning and tell the carriage to wait—unless, which is highly probable, one sleeps aboard....

> And thirdly there will be the motor omnibus, attacking or developing out of the horse omnibus companies and the suburban lines. All this seems fairly safe prophesying.[143]

Two things stand out from this exposition. The first is obvious: Wells's case for the private automobile—as compared to the then dominant mode of travel, the railroad—is almost exactly the case to be made today for the private aircar versus airline travel.

The second point is more subtle. He describes the car as adding independence to "all the small conveniences of first-class railway travel." In other words, it's a perquisite for the rich man on business or vacation. The common folk ride the motor omnibus, as they did the horse omnibus. Wells's notion of the social and economic penetration of automobiles in the 20th century looks a lot like the current status of private jets. The idea that every man could have an automobile is simply missing. This was a notion farther from the mind of the fin de siècle intellectual than time machines or invaders from Mars.

Historian of technology D. S. L. Cardwell makes the point explicitly:

> If we turn to contemporary speculation in order to gain some idea of men's expectations of the technology of their times, we find that in their predictions of the future, or rather their extrapolations of contemporary technological trends as they interpreted them, writers often made shrewd prophecies. Following the inventions of telegraphy and telephony, television could readily be imagined; air travel by heavier-than-air machines (usually powered by steam-engines and

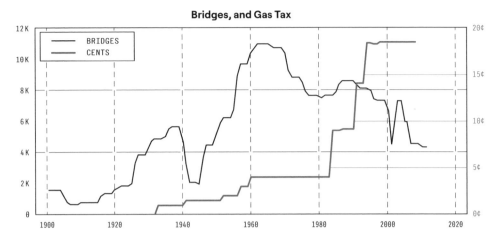

**Figure 23:** Bridge-building in the U.S. peaked in the 1960s; the gas tax rate took off in the '80s.

therefore boasting handsome funnels) was confidently predicted long before the Wright brothers' first flight. Even the atomic bomb was, it is claimed, forecast by H. G. Wells not very long after the beginnings of subatomic physics.

It is a truly remarkable fact that on the very brink of an economic-technological revolution unparalleled in history no one foresaw the *universal* motor car and all that it was soon to imply. This failure on the part of informed and perceptive men to grasp the significance of what was going on right under their very noses must make us suspicious of all attempts to forecast technological developments even one or two years ahead, much less ten or twenty.[144]

This illuminates our investigation of flying cars, and other promised technology, considerably. Rather than asking why we don't have flying cars, it might be more productive to ask how we ever got the family car in the first place. As Wells and others prophesied, the world did indeed fumble along from compromise to compromise, as it always has. No wise global government of enlightened airmen took over and ushered in the era of technological nirvana. And yet, somehow, we did get the family car.

What did the "informed and perceptive men" miss? In a word: productivity. If it takes 10 workers a year to build a car, your economy is simply incapable of building enough cars for everybody to have one, and a car will have to cost at least 10 years of the average worker's wage. Of course, labor is only one of the factors of production. To it you must add materials, buildings, transportation, production machinery, and so forth. As a rule of thumb, during the 20th century the other factors together cost about as much as the labor. In the first decade of that century, the average worker produced about five cars per year; by the 1920s, he produced 20. It was that factor-of-four jump in productivity, spearheaded by Henry Ford and the assembly line, which made the family car possible.[145]

The automobile itself, and then the highway system, enabled the rise of the suburbs. There's plenty of cheap land out there, 25,000 times cheaper than land in Manhattan, and up until the Stagnation people allowed us to take more and more advantage of it. The trend has stopped, if not reversed. One little-appreciated part of the Stagnation was the decline, not just flatline, of investment in transportation infrastructure, which peaked in the 1960s.

By 1970, the nascent Green religion had gotten enough traction that cars began to be demonized. Burying cars became a part of the original Earth Day celebrations. Regulators swarmed like locusts. In 1974, cars were required to have an ignition interlock that prevented them from starting unless the seatbelts were fastened. The national 55-mile-per-hour speed limit was also a 1974 innovation. This was a good measure of how different (and how anti-car) the opinions of the Eloi Agonistes were from the average American: The interlocks (and 85-mile-per-hour speedometers) were so unpopular that Congress quietly repealed them over the heads of the regulators. Even so, by the end of the '70s there was virtually nothing about a car that was not dictated by regulation. It's clear that if we had had the same planners and regulators in 1910 that we have now, right-thinking people all, we would never have gotten the family car.

## Leveling Up

Hans Rosling was a world health economist and an indefatigable campaigner for a deeper understanding of the world's state of development.[146] He is famous for his TED talks and the Gapminder website. He classified the wealthiness of the world's population into four levels:

> › **Barefoot**. Unable even to afford shoes, they must walk everywhere they go. Income $1 per day. One billion people are at Level 1. (Numbers are rounded for simplicity.)
> › **Bicycle (and shoes)**. The $4 per day that they make doesn't sound like much to you and me, but it is a huge step up from Level 1. There are three billion people at Level 2.
> › The two billion people at Level 3 make $16 a day; a **motorbike** is within their reach.
> › At $64 per day, the one billion people at Level 4 own a **car**.

There are, of course, parallel improvements along other axes as well, and Rosling famously also categorized world wealth according to washing-machine ownership, standards of housing, diet, and infant mortality rates. But we will use transportation as our example.

The miracle of the Industrial Revolution is now easily stated: In 1800, 85 percent of the world's population was at Level 1. Today, only 9 percent is. The average American moved from Level 2 in 1800 to Level 3 in 1900, and then to Level 4 in

2000. Over the past half-century, the bulk of humanity moved up out of Level 1, erasing the rich-poor gap and making the world's wealth distribution roughly bell-shaped.

We can state the Great Stagnation story nearly as simply: There is no Level 5. Another factor of four would put Level 5 at about $100,000 per year (per household), which characterizes about a quarter of the U.S. population now. And yet we still only drive cars.

An easy way to see what is happening is to note that the countries still at Levels 1–3 on the development curve have *not* seen a flatline in their energy use per capita over the past 50 years, but, on the contrary, have continued increasing (or, in many cases, started increasing) on a Henry Adams-like curve, just as we did up until 1970.

This matches our hypothesis of the Maslow hierarchy origin of the Eloi Agonistes quite well. If you are below Level 4, you can readily see how life could be substantially better, and work for that instead of virtue signaling. Conversely, in Europe, substantially Level 4, the Eloi Agonistes have arisen with a vengeance; Green parties and policies loom large in politics; regulations have grown like weeds; and economic miracles are shrinking in the rearview mirror. China's GDP is growing at 5.9 percent (according to the World Bank's figures for 2019) and India's at 4.2 percent; France's at 1.6 percent, and Germany's at 0.6 percent. (The U.S. managed 2.2 percent, close to the world average.)

The failure of the future that we were promised was a distinct and proximate result of the flatline of the Henry Adams Curve, as seen starkly in the graph of expected versus achieved technologies. (See Figure 21.) Once we climbed to the comfort levels of the Maslow hierarchy, the Eloi forgot the essential value of energy while taking its benefits for granted. Eloi Agonistes then had fertile ground to demonize powerful technologies, and did so for status and profit. Bureaucratic regulation and the Machiavelli Effect, always the drag terms in the social dynamics equation, saw their coefficients increase exponentially. Ergophobia became the law of the land and a shibboleth of polite society. And we had a run of "bad luck."

As we shift our focus from past to present, from "What happened?" to "What could have happened?," we shall largely shift from considering stumbling blocks to exploring possibilities; but it is always valuable to keep in mind why we don't have flying cars, don't have power too cheap to meter, don't have nanotechnology. And to gaze pensively at the Moon, where human footprints have lain undisturbed for 50 years.

# II

# Profiles of the Present

# 9

# Ceiling and Visibility Unlimited

It has been feasible since the 1930s to manufacture flying cars that are point-to-point, as least for the country estate crowd and the country club set, and since the '40s we have been able to produce helicopters that can travel from just about anywhere to anywhere. We could have built them, but we didn't. There are many other things we could have built but didn't, ranging from the industrial to the molecular scale. The things that we are fairly certain we could have done *by now* constitute, in some sense, a Failure of Nerve; we will consider that for the present. We shall leave Failures of the Imagination for the future.

But any story of the human use of technology must examine the human element. If people actually had flying cars, could or would they use them? Only about 0.2 percent of Americans are pilots, and only 0.07 percent are private pilots who own their own aircraft—and that is at least 10 times the figure for the rest of the world.

Most every time I bring up the subject of flying cars to an acquaintance, I get the same reaction. There is an initial giggle and reference to *The Jetsons*, and then people tell me how flying cars are impossible. Those who haven't flown in a helicopter say it would be too strenuous or scary. Those who have point out that it's too expensive. But almost everyone ends up claiming that the average person who drives a car couldn't fly a plane. The presumption seems to be that flying a plane is of an entirely different order of complexity and competence than driving a car. They are certain that if everybody had a flying car, there would be crashes galore.

Just how hard, really, is it to fly a plane? An objective answer to that would allow us to estimate whether the average person might actually be able to do it safely. With a basic grasp of the question, we could proceed to elaborate concerning issues such as different types of flying machines and degrees of automatic assistance.

I have 50 years of experience in computers, ranging from programming to processor design to robotics (not to mention a Ph.D. in the subject), which gives me a common sense basis as a futurist in the areas driven by digital technology. But with no real experience of flying, I lacked any seat-of-the-pants foundation for pronouncements about flying machines.

Thus, as part of the research for this book, I determined to become a pilot (and, incidentally, acquire an airplane). Becoming a pilot was a fair amount of work, but it was enormously rewarding. In particular, I know a lot more about what flying a car would actually be like, and have some basis for estimating whether or how much it would be something that the average ground-car driver could do.

Throughout the rest of this book, I tend to analyze air travel as if it were purely a question of time or economic productivity, or the like. But such things don't come close to capturing the full value of flight. Flying has been a dream of mankind since the myths of Daedalus and Icarus and the sketches of Leonardo da Vinci. This world is a wonderful place, and you simply don't see much of it from the ground. Your point of view is terribly limited crawling around on the surface.

The very first thing that you notice when you walk out onto the apron at a small general aviation (GA) airport is the feeling that you have been transported to Cuba, where most of the cars are much-maintained models from the 1950s. The same is true of GA aircraft in the United States today. The plane that I bought, a 1977 Beechcraft, is about average. When did you last ride in a 1977 automobile?

The experience of having fallen through a time warp continues when you get into the airplane. You might have the very latest electronic noise-canceling aviation head-set, but you plug it into the radio with two old-fashioned quarter-inch phono plugs. (They are actually slightly different sizes, so they have to go in the right sockets.) Your instrument panel is covered with round mechanical "steam gauges." The basic types of these instruments were set by the RAF during World War II, over 70 years ago. Your engine has a carburetor and a choke, and burns leaded gasoline. And once you start learning how to do navigation computations, you'll find that you are expected to use . . . a slide rule.[147]

Furthermore, the units of measure in aviation are a dog's breakfast. Speeds are measured in knots (nautical miles per hour), but visibilities are measured in statute miles. Altitudes are in feet, sometimes measured above sea level (MSL) and some-times above the actual ground (AGL). Your altimeter reads MSL, but the tower re-ports cloud heights in AGL—except when they are given in flight levels, which are the number of hundreds of feet above an imaginary surface that is where sea level would be according to an altimeter set to a barometer reading of 29.92 inches of mercury. Climb rates are measured in hundreds of feet per minute. Runway lengths are also in hundreds of feet. Temperature and dew point are given in Celsius; dates and times are those of Greenwich, England. Course bearings are measured in degrees, but not from true north; they are measured from magnetic north, which varies depending on where you are. When there is other traffic to watch out for, Air Traffic Control (ATC) reports the bearings in "o'clock" relative to your present course.

Besides the carefully kept-up fleet of planes that would grace any antique car show, you will also find at your friendly neighborhood GA airport a remarkable num-ber of homemade airplanes. In any given year, more homebuilts are put into service than factory-built small private aircraft.

One of the more ironic regulatory pathologies that has shaped the world of general aviation is that most of the planes we fly are either 40 years old or homemade—and that we were forced into this choice in the name of safety.

It did not have to be this way. Without the Great Strangulation, and going by the trend line up until about 1980, the GA industry would be shipping on the order of 50,000 planes a year, total planes in the air would be over a million, and on the average they'd all be a lot newer than the ones flying now. And, as I'm sure your experience with automobiles will bear out, they'd be a lot less likely to break down.

In less than four years during World War II (1941-45), the Army Air Forces lost about 15,000 airmen and 14,000 aircraft in 54,000 total aircraft accidents—inside the continental U.S. A thousand more planes simply disappeared while flying across the oceans to theaters of operation. Once the craft got into combat, 23,000 were shot down and another 21,000 were lost to accidents, weather, and so forth. There were 122,000 airmen casualties total.[148]

The wartime way of thinking—get the job done, whatever the cost—served to obscure the real dangers of flying. Consider that in some cases airmen were sent out with as little as one hour of training in the aircraft that they were flying. It is quite likely that this contributed to a false sense of what risks we would accept in the peacetime postwar world, as compared to the benefits of flying.

So, when all is said and done, how dangerous is general aviation? The FAA and pilots' associations make a big deal of flying safe, and thus, paradoxically, they spend a lot of time talking about hazards and dangers and accidents, resulting in an anecdotal impression of a dangerous undertaking. The people who actually know the risks would be my insurance company. The price for my all-hazards airplane insurance, covering everything from hitting a sparrow on approach to crashing into someone's house, is less than $800 per year. In other words, it's comparable to car or home insurance and considerably less than medical.

The leading cause of death among active pilots is . . . motorcycle accidents.

## See How It Flies

*Riddle: What's the main difference between the sports car and the airplane?*

*Answer: If you speed up the sports car to about 75 miles per hour and pull back on the steering wheel, nothing very interesting happens.*

—**John Denker, *See How It Flies***

Perhaps the most amazing thing about airplanes is that it was possible to build them with 1903 technology. A working airplane can be an incredibly simple machine: a structure with a certain shape, which acts as a glider, and a motor attached to an oddly whittled piece of wood that forms the propeller. There need be no transmission between engine and propeller (my plane has none); the only controls that are absolutely

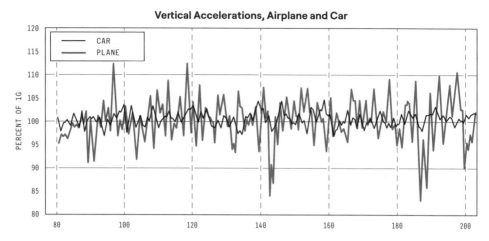

**Figure 24:**  The plane (blue) bounces you up and down more than the car (black) does on a blustery day. Horizontal scale is just seconds into the trip.

necessary can be built by tying wires from a single stick to hinged extensions of the wings and tail. Many ultralight hobby aircraft are built in exactly this way.

So a basic usable airplane is mechanically simpler than a basic usable car. We don't see very many people driving around in cars that they built in their garages.

On the other hand, the plane is definitely harder to drive. We evolved to navigate in two dimensions, not three. Our instincts need some help with going up and down, being tilted sideways, and so forth. We are flat-out terrible at flying when we lose visual contact with the ground: You literally cannot tell which way is up. In fact, the Wright brothers' seminal contribution to flying was solving the problem of control, altitude, and steering. They made bicycles, after all, and understood the importance of balance, and how it tied in with steering.

In my experience, flying is absolutely a learnable skill. As the Wrights' success might lead you to expect, I would say that it is roughly on a par with learning to ride a bicycle. That is, learning to control the airplane is like that. In the air, guiding the airplane is job one of four. The slogan in all the books is "Aviate; Navigate; Communicate." This is a strict order of importance: Make sure the plane stays in the air. Only then make sure that it's going the right way. After that, talk to traffic control and other pilots. And not listed in the slogan is serving as engineer for the motor and other mechanical systems of the plane. Each one of these by itself is not particularly difficult, but doing all four at once is like delivering a really good serve in tennis, or bartending at a busy bar: overwhelming for the novice. After enough practice, though, much of what you're doing becomes habit, routine, and second nature.

As I pointed out earlier, the main difference between a plane and a car is the danger of going slow, particularly when you are trying to land. At the bottom of the landing pattern, you have to make two turns, carefully watching your altitude, attitude, speed, angle of descent, and distance to the runway. And talk on the radio. And watch

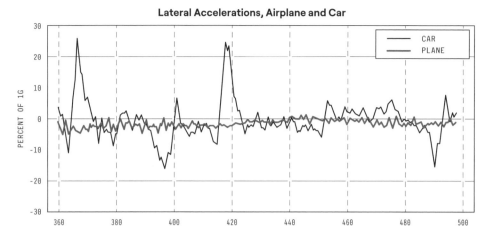

**Figure 25:** But, surprisingly, the acceleration, turning, and particularly braking in a car (still black) are much stronger than the sideways forces you experience in a plane. Note the different scale.

the runway and other planes in the pattern, not to mention birds who get spooked and fly up in front of you. You can't slow down to have more time to cope; in my plane, you have to be doing 80 knots all the way in.

There are plenty of people I wouldn't like to see flying planes. I would guess that perhaps half the people who currently drive cars shouldn't fly planes under manual control. But the bottom line has shifted since 1962. Already there exist automated engine control and monitoring systems, which relieve you of some of the engineer's duties; GPS, which makes navigation much, much easier; and autopilots. There are now fully autonomous aircraft that can do the entire job by themselves; you need only tell them where to go.

To summarize, it's hard but not that hard to learn to fly a plane. Doing so lies within the abilities of, by my rough guess, half of the population currently driving cars. Planes are also as safe and arguably safer than cars. And they are majestic—the experience of looking out an airplane window has no equal in a car. So perhaps the most unexpected aspect of owning an airplane is how many of my friends and acquaintances turn down offered joyrides. Not a huge number, perhaps 25 percent, but there are plenty of people out there who get queasy at the idea of a ride in a small plane. On a blustery day at a fairly low cruising altitude, they have a slight point, but on a calm day they don't.

It turns out that the accelerations you experience in the airplane, even on a bumpy, blustery day, are not greater than you get in a car; they are just different. In particular, the bumps you get are vertical, and slightly more frequent. In a car you are subject to forces just as great, indeed greater, but typically they are sideways—starting, stopping, and turning. (See Figure 25.) It seems likely that if we grew up riding in airplanes, this would be something we'd get used to.

The other problem that people have with small planes, not to mention large jets, is noise. Inside your plane you typically wear a noise-canceling headset. Outside, the

**Airflow at 0° Angle of Attack**

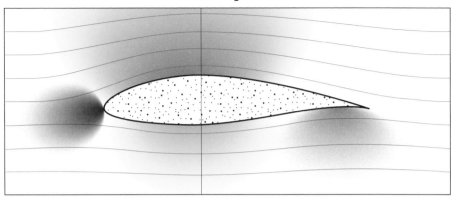

**Figure 26:** A NACA 64(3)-618 wing section at 0° angle of attack, with the flow of the air around it.[151] There is an area of high pressure (black) underneath, and a larger region of low pressure (blue) above it. A wing like this can have a theoretical lift-to-drag ratio of over 100.[152]

plane can be heard a mile away, particularly on takeoff, when the throttle is revved up full. But this is a problem that can be addressed technologically. Ultimately, perhaps the biggest problem that people have with planes, small and large, is that we don't instinctively understand how they work.

## Windy!

It is blowing gale-force winds outside, 35 knots sustained with gusts in the 40s. The house is being pelted with pinecones blowing off the trees; occasionally a branch comes down. Some neighborhood houses lose siding or shingles. You can walk upright in a gale, but you must pay careful attention to each step. You have to lean hard against a door to open it into the wind.

Now imagine that you are in a hurricane-force wind of 70 knots, twice as fast. The forces you feel are not twice those of the gale—they are four times as much. You cannot stand upright; the wind force is similar to your weight. We are accustomed to traveling at 70 knots (about 80 miles per hour), but we rarely experience it directly, and typically we are unaware of how much force is being exerted by the air. At a nice flying-car speed of, say, 350 knots, forces are 100 times that of the gale. The trick is to use them to hold us up in the air, without being blown to a standstill.

Imagine a box, one foot cubed in size, suspended in midair. At sea level, the ambient air presses on each side with a pressure of 14.7 pounds per square inch, or a little over a ton. The box just sits there, however, since the forces balance for a net of zero.[149] Now imagine the box contains a vacuum instead of air, and that the top of the box magically vanishes.

For a millisecond or so, two things will happen. First, the upward force on the bottom of the box will not be balanced by a force on the top. Second, air above the box will accelerate downward into the box.

**Airflow at 6° Angle of Attack**

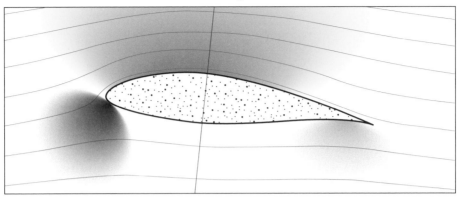

**Figure 27:** The same wing at an angle of attack of 6°. Notice that the high-pressure area where the oncoming wind strikes the leading edge (the "stagnation point") has shifted around to the bottom, and the low-pressure area is deeper, broader, and shifted to the front. Lift coefficient rises from 0.4 to 1.0, meaning you get two and a half times the lift.

Basically, this is how a wing works. The difference is that the air is rushing sideways past the wing, so that when the accelerated air in our gedankenexperiment would have hit the bottom of the box, the air passing over the wing falls behind it and continues downward. There's a conceptual similarity to the reason a satellite stays up: It falls the same as any other object in the Earth's gravitational field, but it has such a sideways velocity that the surface of the planet curves away from it at the same rate it falls. Intuitively, we say that there is a centrifugal force holding the satellite up; we could as readily say that the centrifugal force of the curved path of the air over a wing produces a pressure gradient that maintains a partial vacuum there.[150]

Typical airplane wings derive most of their lift from this lowering of pressure above them, rather than from increasing pressure below them. This is by no means necessary. When the flaps are lowered on a conventional wing, a rise in pressure below is a major part of the lift enhancement. We do this when we are trying to slow down, because it also increases the drag. The reason for the usual design is efficiency: The normal wing has a higher lift-to-drag ratio and thus requires less thrust for a given amount of lift.

The hardest thing to grasp intuitively about any (subsonic) fluid flow pattern is that the pressure changes affect all of the air in all directions—so the incoming air seems to "know" ahead of time that the wing is coming and rises to meet it. What's happening, of course, is that with a high-pressure area under the wing and a low-pressure area over it, there is a pressure gradient pushing the oncoming air up. If you zoom back from either of these closeup views of the wing and flow, the "farfield" flow is exactly level, left to right. The lift force is at right angles to this flow, that is, it is a force exerted straight up. In theory, or in a wind tunnel, the drag force can be as low as 1 percent of the lift. Thus, if the path of the wing through the air is tilted even slightly downward, the lift vector is tilted forward enough to overcome the drag; that's how a glider (or autogyro) works. It is also why a well-designed glider can go 50 feet forward while only falling one foot in altitude.

Lift is proportional to the square of the speed and, within certain limits, to the angle of attack—how tilted the wing is with respect to the incoming airflow ("relative wind"). The thing that makes piloting an art and a skill is that the angle-of-attack relationship fails at an angle somewhere between 15° and 20°. Instead of increasing lift, further tilt causes it to disappear, a phenomenon known as stall.

The word "stall" was used in 1904 by Wilbur Wright in describing the phenomenon, and it stuck. When the Wright brothers first made wings, they modeled the shape on birds' wings—thin, with a concave undersurface. These can have high lift, but they must be oriented very precisely to the oncoming air. The bird does this as automatically as you balance on your feet, but for a rigid wing, rigidly attached to a machine, that design was finicky; piloting a Wright airplane required a lot of skill and constant attention. Aviation pioneers through the 1910s modified the wing sections to have more rounded leading edges and flatter bottoms. These are considerably easier to fly.

If you slow your airplane down, reducing the velocity factor of lift, you must pitch the nose up, increasing the lift coefficient factor, to maintain lift equal to the weight of your plane. As you do this, you will begin to notice a buffeting, as if you were driving on a very bumpy road. The speed of the air past the wing is no longer enough to get it past the low-pressure area, which pulls it back in. It begins hitting the back of the wing. Vortices (i.e., "bubbles" of rotating air) form over the trailing edge of the wing. As a vortex forms, it ruins the effective shape of the wing, as far as the airflow around it is concerned, and cuts down the lift. Then the vortex blows off, lift is momentarily restored, and you've felt a bump. This happens several times per second. Your wing is beginning to stall. If you pitch up any more, the flow will not be able to reattach even momentarily and you will lose lift dramatically.

Stall is what makes flying difficult and dangerous. The handbook of your airplane will list a "stall speed," but that is at best shorthand. A wing can stall at any speed; it stalls at a particular angle of attack. The stall speed listed is the speed at which the stall angle of attack generates lift equal to the weight of the airplane. Go slower than that speed, and either the plane stalls or it isn't generating lift equal to its weight. In either case, it goes down.

As my flight instructor told me, "To go up, pull back on the yoke [which increases your angle of attack]. To go down fast, pull back some more."

Which makes another point: Being low to the ground is dangerous, too. If you're high, you can afford to fall several hundred feet while untilting the plane, speeding up, and regaining normal flight configuration. But you can easily fall the height of a 20-story building while doing this. So going low and slow is a particularly dangerous combination. It's no wonder that airplane accidents cluster around takeoff and landing.

Flying high has its own dangers, however. In particular, anoxia can sneak up on you, resulting in a condition that is like being very drunk without realizing it. This is magnified if you have even a little alcohol in your bloodstream; drinking and driving is

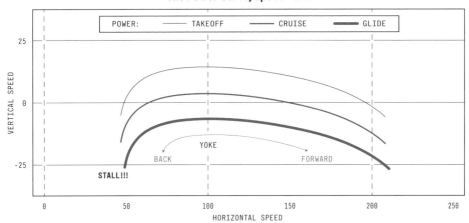

**Figure 28:** The power curve, speed form. Both scales are speeds. Pushing or pulling the yoke (which tilts the nose down or up, respectively) moves you along the curve; pushing or pulling the throttle moves the whole curve up or down.

a bad idea, but drinking and flying is a very bad idea. On a long flight, dehydration can cause some of the same symptoms, as thin air can dry you out a lot more quickly than you realize. A small plane with a normally aspirated (i.e., not turbocharged) engine can have a service ceiling in the neighborhood of 14,000 feet, high enough for you to need supplemental oxygen if you're up there very long.[153]

That going slow is dangerous is quite counterintuitive, given our experience in cars. If I had to list any one thing that was the biggest problem with having every Tom, Dick, and Harry flying airplanes, this would be it. It is surmountable, of course, both by training Tom, Dick, and Harry and by mechanically improving the plane. Still, the way that an airplane reacts to its major controls is counterintuitive, and can be summed up neatly in a diagram that turns the power curve upside down. Use vertical speed instead of horsepower for the vertical scale; each 100 horsepower can raise a 3,300-pound airplane at 10 knots (although your vertical speed dial will call it 10 hundreds of feet per minute). (See Figure 28.)

On the front end of the curve, you push the nose down and go faster, both forward and downward: You're diving, and cashing in your altitude energy as speed. In the middle is a speed that is the best climbing speed with the engine on, and the best gliding speed with it off. At speeds less than this, on the back end of the curve, things work backward: Pulling the nose up makes you slow down and go down, and you're flirting with a stall.

The curve itself shifts up and down as you add or remove power. On a particular airplane, the shape of the curve may change slightly, but that's a secondary effect you can usually ignore. For example, the peak of the curve is the best climb speed with full power *and* the best glide speed with no power. These are typically listed separately for a given airplane but will generally be within a few knots of each other.

One of the less intuitive things about flying is that you typically move the yoke, pointing you up or down, to go faster or slower, while actually going up or down with the throttle. This makes more sense if you look at the intersection of the "cruise" curve and the zero-vertical-speed line. To go up, move the whole curve up, using the throttle. To go faster, move along the curve with the yoke, only adjusting the throttle to maintain altitude.

The other aspect of an airplane that is significantly different from a car is that the airplane is essentially balanced on its wings. The elevator gives you control in pitch, but it is like the toe that you keep in contact with your desk when you lean back and balance on the back two legs of your chair. It will not support you if you are seriously unbalanced. So loading passengers and cargo into the plane is like loading a rowboat: You have to pay attention to where the center of gravity is, and in particular to the total weight. If the gunwales are awash, don't lean out over the side! The closer to the max load you are, the more nearly you must be perfectly balanced.

Aerodynamic forces on an airplane vary with the square of the speed through the air. If you increase the weight, you can increase the speed to make up for it. You glide at the same angle, for example, approaching for a landing, but you move along the same path faster. Which means that the runway that had plenty of room when you were flying by yourself suddenly looks a lot shorter when you have some passengers and luggage on board.

By far the biggest drawback to flying as a means of travel is weather. If you are equipped (and licensed) to fly on instruments—that is, without any visibility out the window at all—and have an airplane that is capable of resisting ice formation, your windows of flying opportunity are somewhat closer to those of commercial airliners. Otherwise, it is not uncommon to get stuck somewhere and just have to wait the weather out. Of course, if instead of a 1977 Beechcraft you had a modern convertible, you could just fold up the wings and hit the road.

Imagine that you're flying along at 1,000 feet, approaching the airport for a landing. You're flying into the wind, which is blowing at 20 knots. You come down to 500 feet and you cross into the sheltered, low-wind, surface air—or worse, into wind that's blowing in the opposite direction. Your airspeed has suddenly dropped by 20 knots, easily enough to make you stall, since you were slowing down to land. This is called wind shear. It can be deadly.

Besides wind, there are other aspects of weather that make flying, not just landing, hazardous. The most common is simply clouds, haze, and fog. Flying into a mountain that you didn't see because your windshield was a blank sheet of opalescent gray will ruin your whole day. Worse than that, if you are in a cloud or even flying over water on an overcast day where haze obscures the horizon, you can't tell if you are really right-side up or upside down. If you're upside down, you feel the same force on the seat of your pants—but it is because the airplane is actually accelerating downward at two G's. This is extremely unintuitive and has caused quite a few accidents, of a type that the FAA euphemistically refers to as "controlled flight into terrain." One of the harder things to learn about flying is to trust your instruments instead of your instincts.

You also have to learn to trust an anonymous voice on the radio that can make the difference between hitting another airplane or not.

## Air Traffic Control

"Smallville Ground, Toyota six five four Echo Bravo Charlie at seven seven three Elm Street, taxi for interstate niner five south with information Golf."

You have gotten into your car and are about to drive to Aunt Millie's house for tea. Since your home is in "controlled groundspace," you cannot simply pull out into the street, but must contact Ground Traffic Control for clearance. Your Toyota's license plate is 654-EBC. You speak in a cryptic code to save radio time, since everyone in town uses the same frequency. You have to verbally identify who you're talking to, who you are, where you are, and what you want to do. You tell the controller that you already have the weather report so he doesn't have to read it to you over the air.

"Toyota Echo Bravo Charlie, Smallville Ground, altimeter two niner niner two, proceed Elm Street south, Nathan Alley west, Fairmont Way south, hold short of Market Street."

You now have right of way up to Market Street, where you must stop. You also have to acknowledge that you heard the stop instruction: "Echo Bravo Charlie, hold short of Market Street." You can add "Roger" if radio traffic is light.

And so you proceed across town to the freeway, asking and getting clearance every few intersections. Finally you switch frequencies to a different controller:

"Smallville Tower, Toyota six five four Echo Bravo Charlie ready for departure at interstate niner five south ramp one six seven Bravo, request left lane."

After a wait to allow a few big trucks to go by, "Toyota Echo Bravo Charlie, cleared for departure, continue right lane until advised."

"Echo Bravo Charlie."

Now you're on the highway and need to take the Middleburg bypass.

"Toyota Echo Bravo Charlie, contact Middleburg Approach on one two three point four five."

The controller for the highway has handed you off to the controller for the bypass, who will direct your further progress. You toggle in yet another frequency on your radio.

"Middleburg Approach, Toyota six five four Echo Bravo Charlie checking in."

"Toyota six five four Echo Bravo Charlie, squawk four three four two and ident."

This is a setting for your radar transponder, which tells the controller your altitude. When you press the ident button on your control panel, your dot on his radar screen blinks.

"Toyota six five four Echo Bravo Charlie, altimeter two niner niner two, maintain course at three thousand."

Your altimeter is essentially a barometer that reads out in feet above sea level. But barometric pressure is always changing, which is why we use it to predict weather. You have to keep yours calibrated to whatever the sea-level pressure would be wherever you are. Air traffic controllers will provide you with the setting you need; if you are flying VFR, i.e., not under a controller, you are well advised to tune to every nearby airport you pass and pick up their altimeter settings.

There is, in fact, quite a lot of airspace where you can fly however and wherever you want (in good weather). The reason for air traffic control, or for using the radio in uncontrolled airspace, is to avoid accidents: collisions or near misses where the wake of a big jet can tumble a light plane out of the sky. The ground controller at a big airport (after which I modeled the town traffic controller above) is necessary because virtually all the runways and taxiways are one-lane roads and airplanes can't back up.

With today's technology, building a capable, efficient flying machine requires a design optimized for flying. Virtually all airplanes are so optimized, the result being that they are ungainly and slow on the ground. They require wide taxiways with relatively gentle turns. They are dangerous to be around: Besides the propeller, the air blast will take any loose pebble on the pavement and throw it backward as hard as you could have thrown it by hand. Airplanes can't back up under their own power (they don't even have rearview mirrors); to back into a parking space or hangar, the pilot has to stop the engine, hop out, and push the plane (by hand or by tractor). That's one of the reasons for ground controllers at busy airports: Most taxiways are one-way-at-a-time and the traffic needs direction from someone in a tower who can see everything at once.

But the main reason all that radiocom is necessary is that airplanes in the air, unlike Toyotas on Market Street, are surprisingly hard to see. This is unintuitive, because even a small plane is a fairly large object in everyday terms—it's generally about 30 feet wide. But the plane that you're looking for is typically miles away. When you watch out for traffic on the road, you're looking for a car that may be 100 feet away. The area of the airplane as a spot on your retina could be 10,000 times smaller than the car. What's more, the plane can be anywhere against a cluttered background, whereas the car will appear in a relatively circumscribed, predictable area.

Add to that the fact that you might have some clouds, that the far distance is usually hazy even on an apparently clear day, and that closing speeds can be several hundred miles per hour, and you can see why being on the radio is a good idea.

"Toyota six five four Echo Bravo Charlie: radar service terminated; squawk VFR. Frequency change approved. Good day!"

# The Hundred-Dollar Hamburger

*For most gulls, it is not flying that matters, but eating. For this gull, though, it was not eating that mattered, but flight.*

—**Richard Bach,** *Jonathan Livingston Seagull*

Jonathan L. to the contrary notwithstanding, it turns out that quite a few smaller airports make an effort to be home to pretty good restaurants.

It is 52 nautical miles, as the crow flies, from the Accomack County, Virginia, airport to the general aviation airport at Colonial Williamsburg. My airplane does that at normal cruising speed in something under half an hour. It is a pleasant excursion on a nice day to fly to Williamsburg, have lunch, and fly back. Driving to Williamsburg takes more than two hours each way, including crossing the Chesapeake Bay Bridge/ Tunnel ($18 toll). A little farther from Accomack County, in the opposite direction from Williamsburg, is the Georgetown, Delaware, airport. Both of these, and several others in the same general area around the bay, are nice lunch destinations because they have decent restaurants, catering to private pilots, right in the airport. No need to rent a car after you land.

The phrase "hundred-dollar hamburger" is common among pilots and aviation writers. Like many such wry clichés, it has more than one meaning. The most obvious is saying, with an implied wink, that I went out flying to get a hamburger, when we all know that I went out flying because I like to go flying, and the hamburger was an excuse. Or I might be expressing annoyance that the best excuse I can find for flying is to go out to lunch. Or to push to the far end of the spectrum, I am seriously disgruntled that general aviation today is so circumscribed as to be essentially a very expensive hobby and not, at least not nearly as much as it ought to be, a useful mode of transportation.

We pilots might gripe about hundred-dollar hamburgers, but commercial aviation is far worse. Imagine that you were trying to take an airline flight to make a lunch date. The effort requires reservations, ticketing, security lines and X-ray scans to make sure that you aren't accidentally carrying a pocketknife or bottle of water, gate changes, waiting in line at the gate in Boarding Group 7, waiting to disembark while everyone in front of you struggles with their carry-ons (don't even bother with checked luggage), getting a rental car, and arriving on time. And then repeating all of that to get home. If you tried to fly to lunch commercially, even if the actual flight itself were instantaneous, you would do nothing else that day.

Aviation as a whole has a major last-mile problem. From the point of view of general aviation, "so near but yet so far" describes it best. However, with the destination restaurant right there at the airport, the hundred-dollar hamburger gives us a taste, so to speak, of what the world could be like with flying cars.

# 10

# Dialogue Concerning the Two Great Systems of the World

The history and the technology of flying cars are both interesting, although for different reasons. The question of whether people could actually fly them took more answering than I expected, but the answer is a qualified yes: A substantial number of people could pilot the kind of flying cars that were built in the '50s, and certainly could handle modern craft with automated assistance. But the question then shifts to how much effort people would be willing to put in to do so. We know that flying cars could be—have been—built. Now we come to the hard part: How much would it be worth to you to have one?

Ever since "Green Light for the Age of Miracles," the 1944 *Saturday Review* article, told us that there might be cars with wings and helicopter flivvers in the postwar world, the notion of the flying car has been bifurcated into two major categories, which we will refer to as convertibles and VTOLs.

Headlining the latter category is, of course, the helicopter. In New Zealand, the helicopter's advantage is magnified by rugged terrain, sparse roads, and windy air. New Zealand has one helicopter for every 6,000 people. We in the U.S. have only one helicopter for every 35,000 people.

The popular Robinson R44 has about the same total weight, interior size, useful load, cruise speed, and operational ceiling as my airplane. The two vehicles have similar engines, both examples of the workhorse Lycoming air-cooled horizontally opposed aero engine, but mine is a four-cylinder while the R44 sports a six. The only big difference, flying, is that my fixed-wing airplane has about twice the range of the helicopter.

The real difference, of course, isn't in flying but in taking off and landing. Instead of thousands of feet of runway, 50 to 100 feet wide, the helicopter only needs a pad that is 10 feet by 10 feet. The amount of land taken up is little enough that most hospitals—and many high-end resorts, corporate headquarters, and snooty wineries—have helipads, even though the amount of traffic is typically nominal.

It isn't just that they take up so much less space. Taking off in a helicopter is a much more civilized proposition than in an airplane. Instead of taxiing out to the end

**Figure 29**:   The popular Robinson R44 helicopter.

of a runway, lining up, shoving the throttle to max, and careening down the runway, you simply sit in your helicopter while the motor runs the rotor up to speed. Then you tilt the wings for lift, just as you do in the airplane—except that in the airplane this involves tilting the entire airplane, but in the helicopter you only need to tilt the rotor blades. So your seat just moves smoothly up, like some improbable stair lift. The vehicle accelerates once it's in the air, but that can be as fast or slow as you like.

There is one other advantage to a helicopter, which also is true of a gyro. It is much less susceptible to the vagaries of gusty winds. Even though you, in the cabin, are moving slowly, over your head the rotor is moving fast, essentially at airliner speeds. The variation in wind speed caused by a gust is a much smaller fraction of the rotor's speed. And because the rotor is moving faster, it can have a lot less area than a slow wing in order to hold up the same weight. And that's a lot less area for the gust to impact. So helicopters (and gyros) can operate easily in conditions that would be difficult, dangerous, and stomach-churning for a small plane.

That is, the helicopter is easier for the passenger. To the pilot, flying a helicopter is somewhat more difficult. The energy chain, engine–speed–altitude, has one more link, rotor speed, and the relationships between the links are more complicated. Controlling the helicopter is much more demanding: The machine as a whole is unstable, the gyroscopic dynamics of the rotor are nonintuitive, and there is a lag between control input and response. In an airplane, you can take your hands off the controls and the plane will more or less keep doing what it was doing. Flying a helicopter is more like standing on a beach ball, especially when hovering.

An interesting bit of history illustrates this. The pilot's seat in an airplane is the left one, like the driver's seat in a car. But the pilot's seat in a helicopter (or gyroplane) is typically on the right. Both Pitcairn's X R-9 and modern helicopters are fairly stable and require only a light touch on the stick (the beach-ball-balancing control). But the first widely used helicopter, the Sikorsky R-4, required a constant and somewhat tiring death-grip on the control. So the pilots in the R-4 preferred to sit in the right seat to use their right hands on the (one-per-seat) stick, leaving the left hand to operate the (shared) between-seats collective control, the radio, and so forth. So a tradition started that has held to this day.[154]

Additionally, the helicopter is mechanically much more complex than an airplane. A simple airplane is just a glider, an engine, and a fan blade. The rotor blades on the helicopter are typically individually hinged, flapped, and tilted; the pilot has a collective control, which causes all of them to increase angle of attack, and a cyclic control, which increases angle of attack on one side and reduces it on the other, in any direction and to any desired degree. The blades perform all these adjustments while rotating and exerting enormous centrifugal forces on the hub. What's more, there is a secondary driveshaft from the engine that runs the tail rotor, and it too has collective pitch, controlled by foot pedals.

So, it is not surprising that a helicopter needs a more powerful engine, more piloting skill, and a lot more maintenance, and that it costs more than an airplane. But the advantages are such that it was not at all unreasonable for the science fiction writers to have imagined helicopters developing into flying cars, especially back in a day when a smaller fraction of the population drove cars and a larger fraction held pilot's licenses. We can get something of a handle on the balance of technical to economic factors by noting that the sector of our economy most flush with cash, health care, has substantially higher helicopter use than the rest. Virtually every hospital has a heliport.

Helicopters today are perfectly capable of the quantum jump in convenience that we would like in our flying cars. Perhaps the most unexpected of the revelations I had while researching this book was how close the speed and load capabilities of a small helicopter are to those of a small airplane. By far the major drawback to a helicopter is cost; the difficulty of piloting one is a distant second.

Today, new helicopters are in the $1 million–$10 million range, and used ones can be had for $100,000–$1,000,000. This is similar to the range for private jets; small piston planes are five times cheaper, new and used. Aircraft engines are quite expensive to begin with; they have to be high-power and highly reliable while still being low in weight. Helicopter engines are the same, only more so. A new Lycoming IO-540-AE1A5 engine, as used in the Robinson R44, lists at $118,481. Most helicopters simply use turbines, putting them in the same class, pricewise, as jets. (The piston-powered Robinsons are in the $300,000–$400,000 range; the Guimbal Cabri G2, Robinson's main competitor for the small heli market, lists for $410,000. You might think of these as being to the turbines what a Vespa is to a Mercedes.)

What you pay for the VTOL capability of the helicopter isn't just money; it's an engineering trade-off that includes low top speed and high fuel costs per mile. Put the same engine you find in an R44 in, say, a Mooney M20, and you get an airplane that can do 250 knots instead of 100.

Technologically, the helicopter speed limit works like this: When the helicopter is hovering, the rotor blades are moving at the same speed through the air, and all you have to worry about is keeping the blade tips subsonic; going supersonic would create a huge amount of noise and shockwave disturbance, and it would waste power copiously. Once the helicopter is in motion, however, the blade on one side is moving forward; this advancing blade has the vehicle's speed added to its rotational airspeed. The opposite blade, the retreating one, has the vehicle's speed subtracted. The faster you go, the greater the difference. But that puts your aerodynamic properties in a vise: Your advancing blade is pushed against the sound barrier, and your retreating blade is pushed against going too slow to generate lift. So the basic standard small helicopter design is pretty much limited to about 125 knots.

A usable helicopter in the family-car class (e.g., the Robinson R44) needs about 200 horsepower (250 at takeoff). That much engine in an autogyro gets you a slightly faster machine: You are not wasting 10 percent of your power just to counteract torque; all of the engine power goes into thrust. The rotor is less loaded because it is only providing lift instead of lift and propulsion, and so, for the retreating blade, stall sets in later. Experimental gyros get up into the 150-mile-per-hour range.

The gyro, if properly designed and skillfully piloted, can land on a dime, but it needs a short (50-foot) takeoff roll.[155] You could put a 50-foot rooftop pad on the top of most homes, were it not for power lines and trees. Back in the '30s, barnstorming autogyros used to land in, and take off from, football stadiums, with room to do a climbout over the stands.

Gyros that could take off vertically were built in the 1930s, notably Pitcairn's last model, the PA-36 Whirlwing. The rotor was run up to about 150 percent cruising rpm by the pre-rotator, and then collective pitch applied. This could produce a one-time hop of about 20 feet, from which the gyro could fly off normally. The mechanism necessary for this, however, made the machine as complex as a helicopter, and the takeoff was demanding and dangerous.

The simplest way to get vertical takeoff with a gyro, in situations where you can't afford even a 50-foot runway, would be to put in a big stationary fan to produce a 25-knot headwind. You wouldn't need more than about 250 horsepower, i.e., a small truck engine. And you wouldn't need this fan for landing at all, meaning that it wouldn't have to be built to aircraft standards. Landings are mandatory, but takeoffs are optional.

The major technical innovation in gyros since their heyday in the 1930s has been Igor Bensen's invention in the 1950s of the "teeter hub" one-piece rotor. A lot of the complexity, and thus expense, of the rotor hub on both helicopters and autogyros was the fact that each blade had to be individually hinged both up/down and forward/back, and that the hinge had to withstand enormous centrifugal force while remaining extremely precise, with no play or looseness. Bensen realized that if there were only two blades, the hinged excursions of one were just opposite of the other. This includes not only the up/down flapping but the forward/back precession, and indeed the cyclic angle-of-attack variation. The way that this works is ingenious but mechanically simple, so the rotor can be one solid piece without the centrifugal force going through linkages, hinges, and bearings. This substantially lowers the cost and raises the reliability of the gyro.

There has been something of a resurgence of gyros in recent years, using the Bensen-style teeter hub and with a small enough engine to fit in the light-sport category (and be relatively inexpensive). They are fun machines to fly but, with a less powerful pre-rotator, don't have as short a takeoff roll as the Pitcairn gyros. But they can still land on a dime.

The gyro continues to be the compromise between the airplane and the helicopter, along a number of different dimensions, including cost, takeoff roll, flight performance, and training requirements. It can be a lot less expensive, and is easier to make roadable. The economic trade-off would depend on the relative costs of land (or roof helipads) compared to machinery. But as we envision a flying car economy, we see a spectrum, with helicopters at one extreme, roadable airplanes at the other, and gyros in the middle. At the extremes, you will pay as much for a helicopter as you would for an airplane that could go twice as fast, but which could only fly from airports. Is it worth it? What is the value of travel?

## Travel Theory

> *A horse! A horse! My kingdom for a horse!*

> —**Shakespeare,** *Richard III*

A classic convertible plane, such as the Aerocar, wouldn't take you anywhere in less than an hour. It took 15 minutes to convert car to plane, another 15 for plane to car, and you had to budget at least a 15-minute drive to and from the airport. There's your first hour, all of it spent on the ground.

On the other hand, convertibles fit better with already existing infrastructure; if you could buy one now, you could use it immediately. For a VTOL, you would need to put a helipad in your backyard. You'd also need every friend or business you were visiting to put one in, too. On the other hand, helipads are pretty simple: For a start, just clear a 35-foot circle of ground—it doesn't even have to be paved. If there were a clear trend toward VTOL ownership, even if penetration were only 1 percent, homeowners and businesses would start putting them in as status symbols.

Even a slow VTOL wins spectacularly over a car, and is only beat by a convertible for trips of 300 miles or more. It wins, that is, for all short trips of the kind you make now in your car. But what about the kinds of trips that you would make if you actually had a flying car? Wouldn't you make more long trips? A long-standing conundrum in economics called the Jevons Paradox will illuminate the resulting choice.

It's generally considered proper for someone who is environmentally concerned to be in favor of energy efficiency. Green organizations are major supporters of corporate average fuel economy (CAFE) standards for cars with higher gas mileage, and for lightbulbs with a higher number of lumens per watt. But you might be surprised to learn that power and energy companies are in favor of these things, too, because they

**The Jevons Paradox**

$10 ROUND TRIP AT $5 PER GALLON

Figure 30:   The Jevons Paradox, often referred to today as the rebound effect.

know something that the Greens tend not to think about: Historically, the more efficiently energy has been used, the more, not less, total energy has been consumed.

This apparent paradox was first noted by economist William Stanley Jevons in 1856, in his book *The Coal Question*. (See Figure 30.) He noted that when the steam engine became more efficient, thanks to the introduction of James Watt's separate condenser (and many other improvements), the amount of coal used in England increased rather than declined. What this meant was, of course, that people were using the new, efficient—and thus less expensive—Watt engines to do many more things than they had been using the older, inefficient—and thus more expensive—Newcomen engines to do.

Let's take a simple example. Suppose you are a farmer who sells tomatoes. You have a truck that gets 10 miles to the gallon. A gallon of gas costs, say, $5. If you deliver a truckload of tomatoes to a market 10 miles away, the round trip costs you $10. Let us also suppose that $10 is as much as you can spend on gas and break even on your tomatoes. You can sell tomatoes to anyone less than 10 miles away, but not to anyone farther.

Now suppose a truck becomes available that gets 20 miles to the gallon. You can now afford to deliver tomatoes up to 20 miles away for the same amount of gas. But notice: A circle with a 20-mile radius has four times the area of one with a 10-mile radius. All other things being equal, you have four times as many customers for your tomatoes with the new truck. So, although the new truck only uses half the gas for any given specific trip, you are making four times as many trips, *and* each trip is twice as long, on average, as the ones you were making before. The result—the Jevons Paradox—is that you wind up using four times as much gas as you did before!

In practice, of course, other constraints, such as the amount of time or tomatoes you have, will weigh into the equation. But it should be clear that there are situations in which efficiency will tend to increase energy usage. Indeed, that is the centuries-

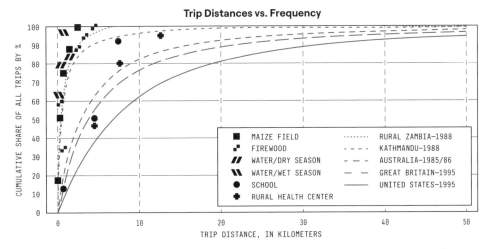

**Figure 31:** Trip distances vs. frequency. Americans make as many 50-kilometer trips in cars as Zambians make three-kilometer trips on foot.

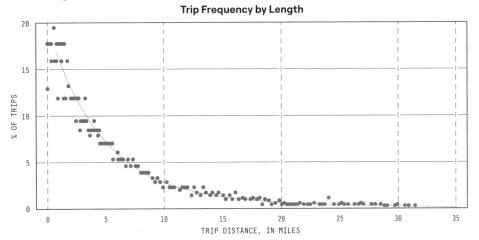

**Figure 32:** Proportions of all trips (from Schäfer's U.S. line) of trips of a given distance.

long historical trend. More to the point, one of the most salient costs of transportation is time; at prevailing wages, speeds, fuel efficiencies, and gas prices, the average American pays 10 cents for gasoline for each mile traveled, but spends 50 cents' worth of time.

So bringing the Jevons Paradox up to date for flying cars means that travel will become cheaper measured in time, rather than coal, yet the result is the same: There will be more travel.

The essential thing to note about the Jevons Paradox is not that people are using more energy; it is that they are getting *more total value* than the mere price reduction would have predicted. More people get tomatoes; the farmer makes more money; so does the gas company. Everybody is better off.

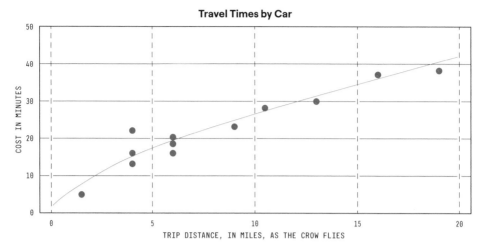

**Figure 33:** The cost in time of trips of various lengths (driven by the author).

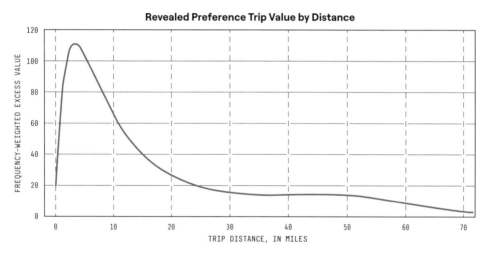

**Figure 34:** The value of trips of various distances to average Americans in cars revealed by the time they are willing to take. The vertical scale is relative value, given that the total of all traveling time is 1.1 hours per day.

There is a research literature in travel theory, which is the study of how much people travel in various environments in different modes. It is mostly used in road and transit planning, but it gives us data and a point of departure. You will not be surprised to find that as long-distance travel becomes more convenient, people do more of it, as Figure 32, from a study by Andreas Schäfer, show.[156]

It is somewhat surprising, but what travel theory allows us to do, to a certain extent, is calculate just how much it would be worth to have flying cars of various types.

Predictably, if flying cars were available, people would make more, longer trips. But the unexpected empirical finding from travel studies is that people in all societies spend about an hour a day traveling, whether they are in Zambia walking barefoot or

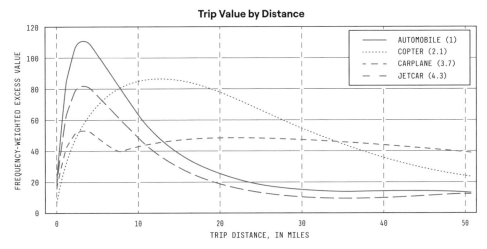

**Figure 35:** The value of the kind of trips you take now in your car, if you had a flying car.

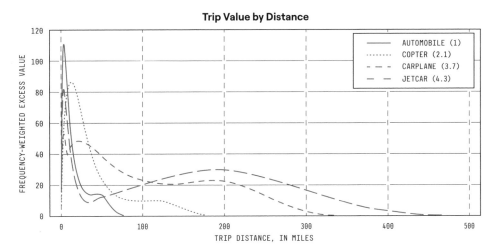

**Figure 36:** The value of the trips you would take in your flying car, if you had a flying car.

in the U.S. riding in an air-conditioned car. Some people travel a lot more than others within a given society, of course, but the average across a given a society is just over an hour per person per day—apparently a human universal. And that allows us to do much more precise predicting than "more and longer."

To get a handle on how much people might travel under situations more advanced than ours, we can fit a curve to the data, so as to extrapolate to longer trips and thus use it in calculations for faster transport modes. (See Figure 33.)

For Americans, this is almost entirely by car. It turns out that there is another more or less surprising universal: Your car does 40 miles per hour. (See Figure 34.) For virtually any trip, of any length, the effective speed of a car, as measured by the time taken to go the point-to-point distance as the crow flies, is 40 miles per hour. You might think that you could do better for a long trip where you spend much more time on the highway traveling fast, but you would be wrong, because the road system is essentially fractal. Highways, on average, take you out of your way by an amount that is proportional to the distance you are trying to go.

Now, the really interesting thing is that we can combine these two functions and derive the value to the average American, as measured in the amount of time they are willing to spend in order to get to whatever destinations there may be at a given distance from where they live. (See Figure 35.)

There are two features of the trip-values graph that have intuitive explanations. The first is the peak in trips under 10 miles. This is due to a combination of the low (time) cost of such trips, and a shadowing effect. If there is a McDonald's five miles from your house, you aren't going to go to one 10 miles away. To the extent that many destinations are alike, the near ones "shadow" the value of the far ones. The other constraint is that you can make two 10-mile trips, out and back, at your effective 40-miles per-hour-speed in your one traveling hour, but not more.

The other interesting feature is the sort of hump going out to about 50 miles. This is probably a day-trip phenomenon, together with the kinds of destinations for which people will make a trip of that length: a ballpark, a hospital, a restaurant fancy enough for an anniversary, etc. These are trips that take most of a day in themselves, but which are taken much more rarely than once a day. In any case, this is the empirically determined value to people of destinations at various distances, as revealed by travelers' willingness to make the trip.

Now let's consider three designs for three flying cars: a helicopter-like one that lands in the driveway, but which can only do 100 knots; a convertible airplane that can do 200 knots and land on a short private strip or straight stretch of road; and a jetcar that can do 400 miles per hour, but which must be flown out of a full-fledged airport to take off.

The next step is to substitute these speeds and overheads back into the value equations and find out how people would travel differently with flying cars of these types, and how much being able to do so would be worth to them. (See Figure 35.) At short distances, the helicopter dominates.

(Note that for these distances, you never actually fly the jetcar!)

This is the obvious advantage of a flying car, and the one most people are thinking of when they imagine a VTOL type: the ability to make the kinds of trips you normally do make, only faster. But the numbers show that that's not where the major value would actually be. (See Figure 36.)

When you look at longer distances, the jetcar dominates. The difference is that you would make a lot more long trips than you do now. Jevons rules. These are higher-value trips, but they're too expensive in time to make very often with a ground car. Note that the value of having the given vehicle, as compared with a car, can be determined by taking the total area under the curves. The reason that the curves for the

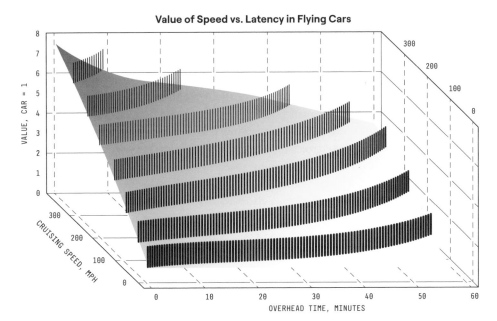

**Figure 37:**   Value of a flying car represented as a multiple of the value of a ground car, as a function of speed and latency (overhead time per trip).

flying cars appear to represent less value for trips less than 20 miles or so is that you'd be taking fewer trips under 20 miles if you had a flying car—you'd be taking longer ones that were of more value to you instead.

We are now in a position to evaluate the value of any flying car, at least as specified by a latency number (the number of minutes added to a trip by having to go to an airport, convert the car to a plane, etc.) and a speed. The result is expressed as a multiple of the value of having a ground car. (See Figure 37.)

A jetcar that was a VTOL, such that you could fly from your driveway and make long trips at 400 miles per hour, would be worth seven times as much as an ordinary car. The jetcar in the previous graphs, the convertible that did 400 but had an overhead of an hour per trip, has about half that value, or 3.5 times that of a car. A fast prop-driven convertible (250 miles per hour) would be about 2.5 times as valuable as a car, and a slow one (100 miles per hour) only about 1.4 times if it had to be flown from airports. That's roughly the same increase in the value of a car that you get from being able to drive to an airport and fly commercially at 400 knots while incurring a three-hour overhead in addition to actual flying time.

In theory, if you would pay today's average of \$35,000 for a car, a fast VTOL jetcar that you could fly from your driveway would be worth \$245,000, as far as pure travel-time value is concerned.

Its value as a status symbol remains to be seen.

# A Compromise

Zimmerman, Custer, de la Cierva, and Pitcairn were right: The inability to fly low, and slow, but even more importantly to take off and land in small spaces, was a major impediment to the general adoption of private flying machines. The helicopter solves many of the problems of the airplane, but it introduces enough mechanical complexity (and a need for more power) that it costs as much as an airplane that travels twice as fast. Even so, I can easily imagine a technological trajectory that would have resulted in 10 percent of American families having a helicopter by now. We could have been following the adoption of the car, just about a century later.

In 1910, the price of a Packard was about $2,500. By introducing mass production, Ford had brought that down to $800 or so with the Model T. By the mid-1920s, you could get a car for $250. A Pitcairn PCA-2 autogyro cost $15,000 in 1931, but by the end of the '30s Pitcairn was selling gyros for under $5,000 (although probably at a loss). It seems not unlikely that with appropriate deregulation and mass production, the private aircraft industry could have been giving us affordable airplanes, and even helicopters, by now.

Seventy-five years after Pitcairn, we should have something at least as good as his 1930s autogyro. Gyros of perfectly usable specs are being built now. This started in Europe, because for years the EU had a light-sport gyro classification that the U.S. did not.

Over the postwar period, the autogyro would very likely have been the technology of choice had we decided, as a society, to go for private flying cars. It would have enabled anyone with an acre lot to have their own landing strip, and it would have enabled many more people to have an acre lot. In fact, in the 1930s Eastern Airlines operated mail-carrying autogyros from the rooftop of the Philadelphia post office. The upward compatibility path to helicopters would have been straightforward when people started being able to afford them and to need that 10-foot helipad.

Travel theory tells us that if we can make a flying car that only has five minutes of overhead and can do 250 knots once in the air, it would be worth about five times as much as a ground car. (Two hundred and fifty knots is a speed limit in U.S. airspace below 10,000 feet, so we will use it as a point of departure for speculation. You can't fly much over 10,000 feet without needing oxygen or pressurization.) What are the chances of building such a thing? How much power do we need, for example, to get 250 knots out of our flying car?

Within the envelope of power-to-weight ratios of 20th-century aircraft, a speed of 250 knots means that we need a horsepower number somewhere between 0.15 and 0.25 times the weight of the aircraft in pounds. Over about 150 knots, you get higher speeds at the same power loading, because you use less and less power to produce a pound of lift the faster you go, so faster airplanes tend to be bigger and heavier. Let's estimate that with high-tech materials we can make the vehicle 2,000 pounds and have it carry 1,000 pounds of payload. That means we need somewhere on the order of 600 horsepower. And that's a problem: A piston engine producing 600 horsepower weighs on the order of 900 pounds. Airplanes with those specs were produced, in the 1930s and '40s, but they were fighters and racers, more than half engine. We would be much better off with a turbine. The Lycoming/Honeywell LTS101-650C, for example, packs 675 horsepower in only 241 pounds.

This makes a huge difference. There's a good reason that "Jet Age" vied with "Space Age" as a description of the era dawning in the 1950s. The turbine, at least as much as any other new technology, was iconic of the future; they were, after all, the "Jetsons" in their jet-powered car, and not the "Spacefields" or even the "Atoms Family." And in fact, over the succeeding half-century, turbines have completely changed the technological landscape, especially in air travel, where their very high power-to-weight ratio is of most value. Another critical but less commonly appreciated value is that they are significantly more reliable. A piston aero engine must be overhauled every 2,000 hours or so, while for a small turbine it's around 10,000. At 250 knots, your turbine will transport you the better part of three million miles. Commercial air travel as we know it today would be impossible without turbines.

Turbines are pricey, but not out of reach given the assumptions of this chapter. Even without assuming away the Great Stagnation, we might imagine that, had mass production and innovation taken their natural courses, a turbine-powered flying car would cost the same as a house. Since about 5 percent of American families have second homes today, there could be plenty of turbines out there, if we had but wanted them. If we had somehow managed to avoid stagnation, and incomes were three times higher than they actually are, flying cars might have reached the levels of adoption that ground cars had in 1920.

The bottom line is straightforward: Futurists in the 1950s and '60s making technologically literate predictions of flying cars assumed that they would use turbines. A turbine would give your flying car the power to take off vertically and land on a dime like a helicopter, or fly at 400 knots as small bizjets do today. Travel theory tells us that either of these would give you a vehicle about three times as valuable as your car, but both together would make it more like seven. The really interesting question is whether that's possible.

Start with the assumption that you require VTOL. Given the technical success of ducted-fan VTOLs, such as the VZ-4 and X22-A, why didn't they catch on as flying cars? To some extent they did. The military, notably the Marines, flew the Harrier VTOL and now fly the V-22 Osprey, a VTOL flying bus that fulfills the roles the X22-A was envisioned for. The F-35B Joint Strike Fighter is a short takeoff and vertical landing (STOVL) craft, capable of taking off from, and landing on, ships much smaller than a full-size aircraft carrier. But there is always a trade-off. The vast majority of aircraft that are fast are not VTOL, and ones that are VTOL are not fast.

One reason is power. Turbines use a lot of fuel, but only because they're producing a lot of power. A piston-engine plane at 100 knots might use 10 gallons per hour (mine does), whereas the recently certified Cirrus Vision "personal jet" flies at 300 knots us-

**4,000 lbs. Thrust**

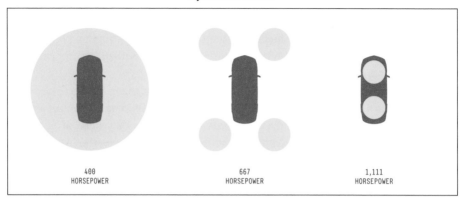

Figure 38: Helicopter, quadcopter, and VTOL arrangements of thrust fans shown against the footprint of a Toyota Camry.

ing about 150 gallons per hour. But the power requirements go up as the cube of the speed; everything else being equal, you'd expect to use 27 times as much fuel (per hour) to go three times as fast. If you're going to put that much engine in your VTOL flying machine, you're buying a private jet—but one that can only go half as fast as a private jet.

A car-size craft with 1,000 horsepower should be able to do airliner speeds, e.g., 450 knots. If, instead, you go slow and use the power to hold yourself up, fuel per mile goes up alarmingly. At the extreme, stationary hovering, fuel per mile is infinite! Alternatively, you could put the same 1,000-horsepower turbine in a helicopter (e.g., the Bell UH-1 "Huey") and carry four times the payload—but only cruise at 110 knots. So there really is a design dilemma: Other things being equal, being a VTOL costs you half your speed as a flying machine.

Suppose, for example, that you want to build a VTOL flying car by having four 10-foot propellers deployed into a standard quadcopter arrangement around the car. They would have a total disc area of 314 square feet. The total weight of your car is 3,000 pounds, and the rule of thumb is you need twice the weight of the vehicle in thrust for safe takeoffs and landings. That gives you a disc loading of about 19 pounds per square foot. You can lift six pounds per horsepower, so you need 1,000 horses.

In practice, helicopters use more thruster area and don't get designed to quadcopter peak power guidelines. The reasons for the design trade-offs are mostly due to the difference between the mechanically driven rotor and electrically powered props. The rotor is big and heavy, and acts as an energy store. You spool that up before take-off and draw on it using the collective pitch control. The quadrotor's props have no collective, and must be light for control responsiveness; the vehicle has no energy store equivalent to the rotor. Thus, it needs to have the extra power available in the motors. The extra power headroom is also valuable since, unlike the helicopter, the quadcopter uses rapid differential power variations for attitude control.

**Figure 39:**   The Ryan XV-5B, in NASA livery. (NASA)

In any case, the trade-off is clear. If you want to go fast, you have to squeeze more power into a smaller, faster jet—and get less lifting force out of it. (See Figure 38.)

There are three major methods that have been tried to solve this dilemma. The first is simply to pour on the power and use jets, inefficient as they are at low speeds, for lift. The Harrier is the classic example of this approach. Its engine produces 23,800 pounds of thrust. With a specific fuel consumption of 0.76, it burns 300 pounds of jet fuel *per minute* in hover. This doesn't sound very promising for your flying car.

The second approach is to compromise by using some sort of prop-rotor or tilting ducted fan for both takeoff and cruise thrust. Efficiency can be enhanced in both modes by the use of variable-pitch propeller blades. The V-22 works this way. The recently announced Joby Aviation VTOL air taxi uses six electrically powered tilting lift fans, and reaches 200 miles per hour in the air. Since the Joby is flying now, we may be certain it would have been possible in a non-stagnated present!

The third approach is to use separate lift-off and cruise fans. The main difficulty is that the lift-off fans are going to be big, unwieldy, high-drag protuberances. You're left either having to overcome their drag or retracting them somehow, a nontrivial engineering challenge.

There is one place you can hide the lift fans while in forward flight: inside the wings. An experimental design of this form, the Ryan XV-5, was tried in the mid-1960s. It was supported in vertical takeoff and landing by a downward-blowing fan in each wing and one in the nose. These were covered by lids to form a smooth wing surface while in forward flight. (See Figure 39.)

The XV-5 occupies the top spot in our travel-theory landscape, at seven times the value of an automobile. What's not clear at all is whether it could be manufactured for seven times the price of a car, especially since it would have to be pressurized.

A jet-powered convertible might possibly exist, in a world without the Great Stagnation. It would fit right into today's airspace system; it would fit right into our road system as well. A convertible's best value added stems not from short trips or joy-rides but from longer trips at high speed. A convertible that did 300 knots (like the newly certified Cirrus small jet) but had to be driven 20 minutes to an airport would be worth five times as much as a car. At that speed, from my home in Virginia I could get to Orlando, Memphis, Chicago, or Montreal in about two hours; everything between Boston, Cincinnati, and Charleston, South Carolina, is less than an hour and a half. That is the travel regime in which the flying car would make a huge difference.

I would love to have such a vehicle, and would make quite a lot of use of it. But I fear there are quite a few problems that stand in the way of its being the model flying car for everyone, even in an unstagnated present.

First, could everyone fly a jet? By all accounts, a jet is a bit easier to fly than a propeller plane of similar size; no propeller torque to correct for and no lift effects when you shift power levels. (N.B.: We are not talking about fighter jets!) But there is one elephant in the room: If you are flying a jet, you have to think a *lot* further ahead. We've noted how humans are not instinctively equipped to maneuver in three dimensions. We are not instinctively equipped to maneuver at 300 knots, either. I've noted how landing my plane is like parallel parking at 80 miles per hour. Landing a jet is like parallel parking at 150 miles per hour.

The second major problem with the jetcar is airport congestion. If everyone is trying to go through the airport, backups both on the ground and in the air will increase the latency of jetcar travel, and could easily drop the value of the jetcar from five times as much as a car to three.

And even if you're the only one with a jetcar, remember that for almost all of the trips of the kind you make now, you wouldn't fly. Maybe you should get a helicopter after all.

Or perhaps a gyro. A rotorcraft (either kind) has significant advantages over a fixed-wing airplane in the car-size private-plane category. First and most obvious is the greater number of places they can land. Somewhat less obvious, but just as important, is how often they can fly, because of the weather. Gusty winds are a pain for a fixed-wing: They make flying uncomfortable and landing demanding and dangerous. Foggy conditions are likewise nerve-racking and dangerous. Rotorcraft are much better in both cases than a fixed-wing.

It turns out that with technology very reasonably predictable in a non-stagnated version of today, we really could have a gyro that does almost everything a helicopter does for a considerably lower cost. At a guess, it would look a bit like a drone quadcopter, but optimized for forward flight in one direction, with pusher propeller(s) on the back. Today's rotorcraft have lead bars in their rotors, to provide momentum as an energy store. Remove that but add electric motors in the hubs. With four light rotors, you get some of the advantages that make the quadcopter an optimal design in its class. This gives the hubs separate, or appropriately mixed, tilt control. With the light blades they will adjust to gusty winds extremely quickly for a much smoother ride.

Because it's an autogyro and not a true VTOL, you can get away with about 250 horsepower in the engine compartment instead of 1,000. You could even get away with a piston engine, though a turbine generator would be nicer.

Enhancing autorotation with motors makes takeoff easier and shorter. Landing is already easier and takes no room at all. A 50-foot circle is all you need; it could even be your roof.

But perhaps the best effect of the motors is to make flight more efficient and thus faster, by substituting motor power for some of the autorotation and letting the rotor "sail closer to the wind." Our quad-gyro might very well do 150 miles per hour in the air.

As well as I can figure, that means that, using current technology, we could build a flying car for about three times the price of a high-end automobile—and that flying car would be *worth* three times the automobile.

Further upscale, multi-fan tilt-rotors, such as the recently revealed Joby, weigh in at about four times the value of a car. There are two major provisos that need to be met before these craft can completely supersede helicopters, however. First is the issue of their stability in gusty wind, and, frankly, I have little idea what their susceptibility on that front is.

Second is the issue of power source and range. Joby claims a 150-mile range for its battery-powered craft, only slightly straining credulity. But the theory graphs reveal that even that cuts off about half the value of the vehicle, a large part of which is in longer trips. (See Figures 35–36, in particular Figure 36.) Landing and recharging certainly isn't going to help! So, to fulfill their true value, multi-fan tilt-rotors are going to have to move to turbine generators in the near term, fuel cells in the middle term, and ultimately, let us hope, they will not have to depend on chemicals at all.

# 11

# The Atomic Age

*Should the research worker of the future discover some means of releasing this energy in a form which could be employed, the human race will have at its command powers beyond the dreams of scientific fiction.*

**—F. W. Aston, Nobel lecture**

*The appliances of 2014 will have no electric cords, of course, for they will be powered by long-lived batteries running on radioisotopes. The isotopes will not be expensive for they will be byproducts of the fission-power plants which, by 2014, will be supplying well over half the power needs of humanity.*

**—Isaac Asimov**

The latter half of the 20th century leaned heavily to information processing—computers and Moore's Law and communications and global networks. And the claim is often made that this was because there was a lot more technological headroom for information technology. For the past 50 years, there has been no practical limit to the density, speed, and efficiency of information manipulation, and increasing cleverness has driven the state of the art in an unprecedented trajectory. To this claim is often added the assertion that the same wasn't true of heavy duty, energy-using transportation and construction sorts of stuff, which characterized the first half of the century. Moving matter around requires a certain amount of energy, the argument goes, and energy is expensive. Some people will go further and argue that the optimism and progress of the first half of the century was due to the rapidly expanding supply of cheap fossil fuel, and we ran out of that.

It is true that the optimism and progress of the first half of the century was due to the rapidly expanding supply of energy, as was the optimism and progress of the Victorian era, and, indeed, of the Industrial Revolution as a whole. This is the Henry Adams Curve in a nutshell. But it has nothing per se to do with the source of the energy. We had already shifted from wood to coal, and from coal to oil, as our major energy source. George Jetson's car isn't going to fold up into a briefcase if its tanks are carrying 1,000 pounds of jet fuel.

The ergophobic response of the Eloi is right out of Aesop: Those grapes were sour anyway. Be happy with your 140 characters.

As Nobel laureate F. W. Aston pointed out, this was not the dream of science fiction. From H. G. Wells to E. E. Smith to Heinlein to Asimov, all of the technically literate SF writers knew where the next great source of energy was to be found: We were poised at the dawn of the Atomic Age.

While you are boarding that 747-400 for your vacation in Australia, the ground crew will be filling its tanks with 57,285 gallons of jet fuel.[157] If you, like the average American, use about 500 gallons of gas in your car per year, that's enough fuel to last you 114 years. That amount of Jet A fuel weighs 194.6 tons, and the plane has to carry

**Energy Costs**

| FUEL | UNIT PRICE | AMOUNT/TJ | $/TJ |
|------|-----------|-----------|------|
| Gasoline | $3.37/gal | 7,780 gal | 26,000.00 |
| Electricity | $0.12/kWh | 278,000 kWh | 33,000.00 |
| Natural gas | $12.50/mcf [159] | 948 mcf | 11,850.00 |
| Coal (anthracite) | $170.00/ton | 91,600 lbs | 7,786.00 |
| Tritium | $30,000.00/g | 3.77 lbs | 51,000,000.00 |
| Uranium (as U2O8) | $58.24/lb | 0.028 lbs | 1.40 |
| Uranium (enriched) | $1.63/g | 12.58 g | 20.00 |
| Thorium (oxide 99.9%) | $0.20/g | 12.58 g | 2.50 |
| Boron-11 (p-B fusion) | $5.00/g | 15 g | 76.00 |
| Lithium-7 (carbonate) | $0.06/g | 7.4 g | 0.44 |
| Deuterium (heavy water) | $600.00/liter | 7.8 ml | 4.00 |

Figure 40: A terajoule is the amount of energy that the average American uses in all forms, in about three years.

it, on average, for half the flight—but, in particular, the 747 has to take off and climb to altitude with all of it. In other words, if it didn't have to carry the fuel, the plane could carry nearly 200 tons more passengers and/or cargo. Call it 1,946 people and their carry-ons; compare that to the typical 500 passengers that 747s actually carry. The energy produced by burning that much Jet A fuel comes to 7.5 terajoules (TJ). The same amount of energy could be had by fissioning 94.3 grams (3.3 ounces, or a third of a cubic inch) of uranium.

As I write, the price for Jet A at a nearby airport is $6 per gallon. Fueling up a 747 costs $343,710. Also as I write, the commodity price of uranium, as yellowcake (U3O8), is $49.50 per pound. Only 84.8 percent of yellowcake by weight is uranium, so the effective price is $58.24 per pound, meaning that 7.5 TJ would come to $8.66.[158]

A terajoule is the amount of energy that the average American uses in all forms (including your share of manufacturing, shipping, and worldwide military operations) in about three years. In Figure 40, the prices for gasoline, electricity, gas, and coal are consumer retail; the rest are commodity prices. (They just don't seem to carry uranium at the local Walmart.)

Tritium is a completely synthetic isotope and is currently one of the most expensive substances that exists. I include it for comparison: Used as a beta emitter, it falls somewhere between chemical and nuclear forms of energy storage. Electricity, of course, isn't a fuel, but is also included for comparison. At a rough average, consumer retail energy runs about five times the commodity price of the raw fuel. (The exception is that gasoline costs less than twice the price of the crude oil it's made from.)[160]

Most electricity in the U.S. comes from coal or natural gas. With uranium, since it involves a costly isotopic enrichment step, the markup is more like a factor of 30. Even so, the cost of the uranium going into power generation is trivial compared to the other costs. We could reduce even that by a factor of 10 by going to a thorium fuel cycle, but uranium is already so cheap, on a per-kilowatt-hour basis, that it's not worth the bother.

The average price of a year's energy, in chemical fuels: \$6,553. In nuclear: \$5.80.[161]

## Too Cheap to Meter

At the dawn of the Atomic Age, some reasonably well-informed people imagined that when the technology was mature, nuclear-derived electricity might be "too cheap to meter." Of course, "too cheap" is, in some sense, verbal sloppiness. It's not the absolute price of something that determines whether it makes sense to meter it, but the marginal price—what it costs you to generate one more kilowatt-hour, given the fact that you've already built all the capital equipment and covered all the fixed overhead costs. This doesn't mean that it's free, or even cheap—it just means that you pay a flat fee, as you do for local phone service or basic cable.

But whether metered or not, if you just considered the raw fuel cost of uranium, energy from fission could be very cheap indeed. The cost of actual electricity delivered to your home is due to the cost of the capital—the reactor and generating plant, but also the transmission lines, substations, power grid control centers, and so forth—and overhead, including maintenance and a truly staggering regulatory burden, which multiplies the cost by an order of magnitude. From this point of view, the cost structure of nuclear power is much more like that of solar or wind power than that of fossil fuels. The cost of "renewables" is essentially all capital and overhead.

Let us look again at the history of the Henry Adams Curve, but this time a little more closely. Over the past century, several stages in energy use have followed on each other as we shifted from one source to another. In the 1800s, wood was the major fuel; by 1900, as Adams noted, coal had taken off exponentially. As the 20th century progressed, this was replaced by oil and natural gas.

Nuclear power took off from about 1960 to 1975. Did it have the potential to be the next dominant energy source?

Indeed it did. If it had only followed the same growth curve that gas did from 1920 to 1970, by now nuclear would be producing essentially all of our electrical power. The rule of thumb is that nuclear fuels produce 1 million to 10 million times the energy, per weight, of chemical ones—and thus require the extraction of a million times less raw material, and the production of a million times less ash, than fossil fuel for the same amount of energy. In short, nuclear fuel costs are essentially trivial compared to fossil. Indeed, a wind turbine uses up more lubricating oil than a nuclear plant uses uranium, per kilowatt-hour generated.[162]

While it is an error-prone activity to speculate on what kind of advances might have been made in nuclear power conversion if a major, sustained effort had been made over the past 50 years, it is also an error-prone activity to get stuck on conventional objections. The most common of these is that fission, and hot fusion, produce high-energy neutrons, which are particularly nasty and require heavy shielding.

That's true, as far as it goes. But the objection implies that anywhere you wanted an engine you would have to put a fission reactor. And that is to commit a complete Failure of Nerve.

It was well understood in the 1960s, perhaps even better than today, that you can create artificially radioactive isotopes of various elements by exposing them to neutrons (e.g., in a reactor). Typically, the isotopes produced this way decay by beta emission, that is, changing the extra neutron into a proton by firing off an energetic electron. Beta radiation can be blocked by a couple of inches of water or a thin sheet of metal (for average energies—the betas from tritium are stopped by a piece of paper). At the very least, you could use the heat to produce power, which is, after all, what we do when we burn fossil fuels. Radioisotopic thermoelectric generators (RTGs) were introduced during the Eisenhower administration; but that is doing things the brute-force, inefficient way.

"Beta rays" are, after all, nothing but energetic electrons. Back in the '70s, there was a betavoltaic battery, the Betacel, produced commercially. Completely safe to handle and with a useful life of up to 10 years, they were used in implantable pacemakers. Technologically, they use the high-energy electrons to create a trail of electron-hole pairs in an appropriate semiconductor device. One of the things that has improved a lot over the past 50 years is our semiconductor technology, so there has been a minor resurgence in betavoltaics in recent years. The best that I've seen so far is a Russian lab prototype using a stack of nickel-63 foils alternating with diamond Schottky-barrier diodes.[163] Calculations indicate that, if optimized, it could have 50 times the energy density of a chemical battery.

With 50 years of experience and experimentation, Isaac Asimov's speculation—"The appliances of 2014 will have no electric cords, of course"—was completely reasonable, given the physics and the rate of technological improvement up until then. We really, really should have had atomic batteries by now. And guess what? Your iPhone would never need charging, and your Tesla would have a range of 3.5 million miles. *It is a possibility.*

The key to analyzing the question of how well reactors could have improved over the last half-century is to see just what kind of headroom they had in their design. Molten salt reactors using thorium and integral fast reactors using uranium-plutonium alloys could achieve a 99 percent fuel burn-up, improving both fuel efficiency and waste production by a couple of orders of magnitude over the 1960s designs that we're still using. Pilot programs showed encouraging progress toward each, with Tennessee's Oak Ridge facility operating a molten salt reactor (but without the full thorium fuel cycle) for a year or so, before the project was canceled by Richard Nixon. As *Fortune* magazine noted in February of 2015, the president had banished a reactor that was virtually meltdown-proof, left comparatively little long-lived waste, made it more difficult to fashion a bomb from the waste, ran at friendlier atmospheric pressure instead of the potentially explosive pressurized environments of conventional reactors, and ran at much higher temperatures, making it more cost-effective as an electricity generator.[164]

Thorium doesn't need the separate enrichment step, and the reactor could burn virtually all of it, rather than less than 1 percent. And the element is three or four times as common as uranium. Proven reserves of uranium could power the globe for 77 years; of thorium for 6,472 years.

In addition to fuel availability, molten salt thorium reactors would have two other major advantages over uranium ones. First, it's much harder to divert the fuel to make nuclear weapons. In fact, that is one of the main reasons thorium reactors were not developed in the 1960s: Governments wanted a source of easily weaponizable uranium and plutonium. The second advantage is that a molten salt reactor can be engineered to be walk-away safe. Existing designs for uranium reactors require continuous active cooling to prevent meltdown. But the fuel in the proposed thorium designs is already molten. If the power fails in a conventional reactor, separate emergency power is needed to run the cooling system. In the molten salt reactor, the design works the opposite way: Cooling, enough to keep a plug of salt frozen, is powered by the reactor itself. If the reactor or generator fails, the plug melts, and the fuel flows into a sump where it's too spread out to maintain a reaction.

In other words, the physics of fission allows for a lot of headroom for us to improve reactors over current practice. At a rough guess, we could ultimately get 100 times the power from each ton of fuel mined as we do now. And remember that nuclear fuel is cheap to begin with.

## Radiophobia

There is as much headroom in physics and engineering for energy as there is in computation; what is stopping us is not lack of technology but lack of will and good sense.

Perhaps the most pernicious of the memetic cancers to infect the public mind in the 1970s was nuclear hysteria. One of the best examples of just how enduring this pathological fear remains was the reaction in the media to the incidents at the Fukushima I power plant in connection with the Tōhoku earthquake and tsunami of 2011.

As a result of the earthquake and tsunami, about 16,000 people died (most drowned), 6,000 were injured, and 2,500 went missing. A quarter million were made homeless. A total of 127,290 buildings were totally destroyed; 272,788 "half collapsed"; and another 747,989 were partially damaged.[165] It was a disaster of epic proportions. As part of all this devastation, the Fukushima power plant was damaged and some radioactive materials were released into the local environment. How much danger did this release add to the overall cataclysm? Zero. No one was killed by radiation exposure, and projections for excess cancer from accumulated exposure are negligible. Ten years after the event, the United Nations Scientific Committee on the Effects of Atomic Radiation (UNSCEAR) released a summary study of the investigations performed in the intervening decade. The report stated, "No adverse health

**Energy Source Mortality Rates**

| SOURCE | DEATHS/TERAWATT-HOUR | REMARKS |
| --- | --- | --- |
| Coal (China) | 280 | Air pollution: 500k/yr |
| Coal (world average) | 170 | |
| Coal (U.S.) | 15 | Over 10k/yr in U.S. |
| Oil | 36 | Think LA smog |
| Biofuels/Biomass | 24 | |
| Natural Gas | 4 | Cleaner than oil |
| Solar | 0.44 | Falls from rooftops |
| Wind | 0.15 | |
| Hydro (China) | 1.4 | Dam bursts |
| Hydro (U.S.) | 0.1 | |
| Nuclear | 0.04 | Including Chernobyl |

**Figure 41:** Relative deadliness of the major energy sources. If you are thinking of splitting wood, not atoms, please remember that lumberjack is the number-one most dangerous profession.

effects among Fukushima residents have been documented that could be directly attributed to radiation exposure," and noted that any further delayed effects "are unlikely to be discernible."[166] Ask yourself: Did the news coverage of the disaster reflect this risk ratio? Of course not. The coverage that I saw was filled with hyperventilating about the power plant, and the thousands of people who died from flooding and collapsing buildings were, at best, a footnote.

It is worth pointing out just how much harm and suffering the nuclear fear industry has caused in human terms. On the order of 1 percent of the evacuees from both Fukushima and the 1986 Chernobyl disaster committed suicide.

But what of the Fukushima plume area? What of those photogenic piles of garbage bags, filled with scraped topsoil and labeled "nuclear waste"? Much ado about nothing. The actual measured dose over most of the release was less than the radiation dose from having a couple of CT scans in a year. That was for 2011, the year immediately following the release; by 2012, even this was mostly gone. Fully in the grip of the radiophobic hysteria, the Japanese government evacuated approximately 100,000 people from the vicinity of Fukushima to "protect" them from radiation levels that were about the same as the natural background in Finland. Some 1,600 of the evacuees died from causes ranging from privation to suicide.

You will never come to understand this from the hysterical media, but the fact is that nuclear power generation is the cleanest and safest form of power ever used on a nontrivial scale. This is true even in the case of the old pressurized or boiling-water plants (i.e., the plants we still use), which were designed in the days when computers were built with discrete transistors and ferrite-core memory.[167]

One example of a valuable and sensible piece of regulation was the Clean Air Act. By reducing coal-fired air pollution, it saved over 50,000 lives per year in the U.S.

This was a very good thing—I remember, at age 12, being the youngest pallbearer for my grandfather, who had died of emphysema.

Two-thirds of all the new generating plants started in 1966 were nuclear.[168] Had the main thrust of the environmental concern acted to enhance rather than stifle the trend, instead of reducing pollution deaths to 10,000 per year by making power more expensive, we could have reduced them (and $CO_2$ emissions) essentially to zero by 1990. And, along the way, we would have made energy considerably cheaper. Surely a well-informed public would insist that we stay the course. And yet, the people who rightly wanted to reduce coal pollution were pushed aside by hysterical Eloi Agonistes activists who completely ignored facts to vigorously propagate the falsehood that nuclear power plants could explode as monstrously huge atom bombs, destroying whole cities with a single accident.

That is not possible. They cannot explode. An explosive chain reaction requires an unmoderated critical mass with 80 percent or more enrichment. Power reactors use a moderated reaction in uranium that is only enriched 3 percent. Yes, they are both fission. But then the detonation of a stick of dynamite and the digestion of a stick of chocolate are both oxidation. And yet polls done in the aftermath of the Three Mile Island accident revealed that the activists' misinformation had done its work: Some 66 percent of Americans believed that a power reactor could explode like an atom bomb, vaporizing a city and covering a state with fallout.

Meanwhile, it is radiophobia, not properly engineered nuclear power, that is the silent unseen killer. Energy poverty is estimated to kill roughly 28,000 U.S. residents annually from cold alone, a toll that falls almost entirely on the poor.[169] In the last 25 years, we could well have saved a million lives. But won't a widely deployed nuclear power sector provide an avenue for rogue states and terrorists to obtain nuclear material for nefarious purposes?

As far as getting bomb-grade enriched uranium, this is wildly exaggerated: Turning the 3 percent enriched uranium used in reactors into the 80 percent enriched used in bombs requires the same machinery and nearly as much processing as refining it from raw natural uranium. The resources and effort needed are at a national scale, and every nation with a propensity to try it is very carefully monitored.

In any case, a shift from big 1960s-era power plants to small modular reactors of more modern design—factory-built, self-contained, and buried in operation—would render them essentially unavailable to bad actors. The stagnation in progress in nuclear technology has made the problem of proliferation worse, not better.

## Speed Limit: 1

Most of us understand that you can drink a glass of wine every day with no problem whatsoever, but that drinking 20 gallons in one go would be a bad idea. The same is true of exposure to radiation: On the beach on a clear day at noon, you will be hit by about one watt of ionizing radiation as the UV part of sunlight. If you get 10 minutes

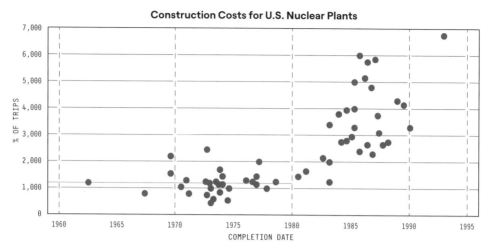

Figure 42: Construction costs for nuclear power plants skyrocketed after the establishment of the Department of Energy. The horizontal line is the pre-1980 average, at $1,175/kW. At today's prime rate of 3.25 percent, that would amortize to about half a cent per kWh. Note that regulation also considerably bloated construction times.

of this a day, you will be healthier than if you didn't, because the exposure causes your skin to produce essential vitamin D. Do this every day for a year, and you're in the pink. But if someone turned a 365-watt UV laser on you for 10 minutes, it would burn you to a crisp.

There is no evidence of a carcinogenic effect in humans for acute irradiation at doses less than 100 millisieverts (mSv) and for protracted irradiation at doses less than 500 mSv.[170] Even the pusillanimous Nuclear Regulatory Commission (NRC) admits, and I quote, "But there are no data to establish a firm link between cancer and doses below about 10,000 mrem (100 mSv)."[171] And they then proceed to point out that this level is 100 times greater than what they allow for public radiation exposure. This is like setting a speed limit of one mile per hour because people have been injured doing 100. You can imagine what regulations made on the basis of this kind of thinking did to the cost of a nuclear power plant.[172] (See Figure 42.)

Yes, nuclear today is expensive. Shipping would be expensive, too, if trucks had to operate with a speed limit of one mile per hour.

A study by economist Peter Lang found that during the 1950s and '60s the cost of nuclear plants was decreasing by about 25 percent for each doubling of capacity. And then the trend utterly reversed. He calculates that, had the earlier trend continued, the price of power would have fallen to 10 percent of what it is as of 2020, and "the extra nuclear generation could have exceeded the actual generation from coal by year 2000 (assuming electricity demand did not change)."[173] And that's not all. "If the extra nuclear generated electricity had substituted for coal and gas generation, about 9.5 million deaths and 174 Gt [gigatons] $CO_2$ may have been avoided."

Nuclear power is arguably the clearest case where regulation clobbered the learning curve. Innovation is strongly suppressed when you're betting a few billion dollars on your ability to get a license to operate the plant. Fewer plants were built and fewer ideas tried. Sherrell Greene, a longtime nuclear technologist at Oak Ridge, described the situation:

> The environment in today's nuclear energy enterprise is hostile to innovation. Not by intent, but in reality nevertheless. The industry is highly regulated. It is very costly to do research, development, and demonstration. It's a very capital-intensive business. The barriers to entry are incredibly high. The downside risks of innovation are more easily rendered in practical terms than the upside gains. Often it seems everyone in the enterprise (federal and private sectors) are so risk-averse that innovation is the last thing on anyone's mind. In this environment, "good enough" is the enemy of "better." Humans learn by failing. It's the way we learn to walk, talk, and ride a bicycle. Our environment today has little tolerance for failures at any level. There's no room for Thomas Edison's approach to innovation in today's world. On top of all of this, or perhaps because of it, the nuclear industry invests less on R&D, as a percentage of gross revenues, than practically every other major industry you might name.[174]

A clear case that bureaucracy was the main factor in choking the advance of nuclear power in the United States can be found in the one place where it didn't happen, the U.S. Navy. Former Navy Secretary John Lehman wrote, "The reason for Navy nuclear success is because there has always been one strong experienced person in charge and accountable, standing like a stone wall against the bureaucratic onslaught."[175] The Navy has over 6,000 reactor-years of accident-free operation. It has built 526 reactor cores (for comparison, there are 99 civilian power reactors in the U.S.), with 86 nuclear-powered vessels in current use.

Remember that, before the bureaucratic morass, construction of nuclear power cost $1,175 per kW. A 2 percent annual increase (i.e., the Henry Adams Curve) in the average American's 10 kW would have required a $235 investment per capita per year. Three doublings since then would have brought us to an all-nuclear grid and put us back on the Henry Adams Curve. Given Peter Lang's estimated learning curve, new capacity would now cost $495 per kW, and increasing the average American's 25 kW consumption at a 2 percent rate would still cost less than $250.

We could have had flying cars in 1940, but for World War II. We could have started on nanotech in 1960. We could have started on solar-powered satellites in the '70s. The low-hanging fruit of nuclear power was left to rot on the ground, and it was sprayed with the kerosene of hysteria and ignorance.

In the 1970s, when the Department of Energy was created and energy per capita flatlined, one heard a lot of well-meaning people decry the space program and say how, instead, we should pay attention to the problems we had right here on Earth. But most of the problems we have here are due to poverty. Poverty is ameliorated by cheap energy. As Bill Gates, perhaps the world's leading philanthropist, puts it, "If you could pick just one thing to lower the price of—to reduce poverty—by far you would pick energy." A survey and analysis of more recent economics publications tends to confirm the relationship. "Energy consumption Granger-causes real GDP per capita and vice versa in the long run, which implies that an increase in energy consumption leads to an increase in economic growth and vice versa."[176]

Counting watts is a far better way to measure a people's standard of living than counting dollars. We do have an economy now that produces twice as many dollars of apparent economic activity per watt of energy consumption, as compared to 1970. But those who claim that is a good thing are missing the point. We have made energy itself more expensive. The number of dollars per kWh generated in the U.S. has increased by 300 percent, while the actual power generated has increased by only 26 percent (more dollars, less value).[177] We have forced everyone to pay more for energy-efficient cars, houses, and appliances (more dollars, less value). We blow hundreds of billions on worthless public transit (more dollars, less value). Yes, we're spending more dollars, but (outside of computing and communication) most of what we're getting for them is virtue signaling. Counting watts, I repeat, is a better way to measure a people's true standard of living than counting dollars.

The Great Stagnation and the Henry Adams flatline were not merely coincidental; they were synonymous.

As Arthur C. Clarke put it:

> In this inconceivably enormous universe, we can never run out of energy or matter. But we can all too easily run out of brains.[178]

He also wrote:

> If, as is perfectly possible, we are short of energy two generations from now, it will be through our own incompetence. We will be like Stone Age men freezing to death on top of a coal bed.[179]

## The Great Physics Stagnation

> Today, education concerning nuclear physics is in an appalling state. On the one hand, discussion of the politics of nuclear issues is to be heard almost daily on the evening news. On the other hand, a visit to any typical bookstore in the U.S. will reveal a total lack of interest in the academic field of nuclear physics. Of course, popular renditions of the "quantum revolution" circa 1920 continue to be written, but even university bookstores rarely have any texts on the nucleus itself. And if you find a book on nuclear physics, check the copyright date! Books written in the 1950s or 1960s and reprinted in unaltered form are on offer to students in the twenty-first century.

> —**Norman Cook,** *Models of the Atomic Nucleus*

Arguably, the most important physical phenomenon discovered in the 20th century was nuclear fission. That was certainly the consensus of mid-century science fiction writers: Several of them, ranging from H. Beam Piper to Larry Niven, dated their future histories from the first sustained chain reaction (or in some cases, bomb). Referring to today's year as 76 AE (Atomic Era) instead of 2018 AD is a subtle way of making the reader feel part of a distinct new phase of history. It's certainly clear that SF writers of the '50s and '60s believed that we were in such a new technological order of things. Outside of science fiction, the existence of nuclear weapons reshaped the dynamics of world politics in ways that we are still grappling with.

So you would imagine that fission would be one of the most studied, and by now best understood, of processes. After all, physics has advanced well beyond it: We have generally accepted theories of quantum electrodynamics (QED) and quantum chromodynamics, which explain the outer atom and the insides of nucleons, respectively. We have county-size particle accelerators, which continue, so far, to confirm the "Standard Model" of high-energy physics.

But, surprisingly, there is still no generally accepted theory that gives a good account of nuclear fission. From the introduction of Kenneth Krane's classic introductory text:

> Nuclear Physics lacks a coherent theoretical foundation that would permit us to analyze and interpret all phenomena in a fundamental way; atomic physics has such a formulation in quantum electrodynamics, which permits calculations of some observable quantities to more than six significant figures. As a result, we must discuss nuclear physics in a phenomenological way, using a different formulation to describe each different type of phenomenon, such as $\alpha$ decay, $\beta$ decay, direct reactions, or fission.[180]

Not too long ago, I had the opportunity to test out Professor Cook's thesis about the current state of nuclear physics literature. One of the lamentable aspects of 21st-century technology is the death of the used bookstore; it had always been a particular pleasure to visit one and spend a couple of interesting hours perusing their stacks when in an unfamiliar city, but the business has largely died. On this occasion, however, I happened to be in Redmond, Washington, home to Microsoft and thus a population of technically interested people (i.e., nerds). A good used bookstore remained in business. Given its clientele, it had a substantial and well-appointed science section.

I visited the store with the express intent to buy every single book that they had on the subject of nuclear physics. This proved distressingly easy. I left carrying three books in one hand, the most current being an overview by Isaac Asimov published in 1991. The interest, demand, and market for the subject had simply evaporated.

Suppose you were going into science or engineering in the 1980s: Would you have picked nuclear physics or nuclear engineering as a promising career choice? Of course not. The number of Ph.D.'s in nuclear physics over the period reflects this clearly. (See Figure 43.)

The bottom line is that interest, and career prospects, in nuclear physics imploded in the 1970s; the field languished and major discoveries stopped coming. A clear

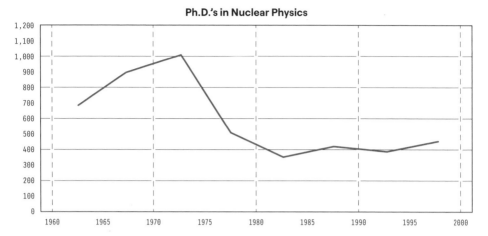

**Ph.D.'s in Nuclear Physics**

Figure 43:  U.S. Ph.D.'s in nuclear physics. (NSF)

consequence is that most reactors still operating today are mid-'60s "Generation II" pressurized water designs.[181]

Not only were there fewer nuclear physicists in the '70s and '80s, but the character of a shrinking field becomes markedly different from that of an expanding one. It's zero sum; politics increases; regulation sets like cement shoes, and the Machiavelli Effect reigns supreme. Paradigm shifts become essentially impossible.

## Clean, Safe, Abundant Energy

*The prospect of cheap fusion energy is the worst thing that could happen to the planet.*

**—Jeremy Rifkin**

*Giving society cheap, abundant energy would be the equivalent of giving an idiot child a machine gun.*

**—Paul Ehrlich**

*It would be little short of disastrous for us to discover a source of clean, cheap, abundant energy, because of what we might do with it.*

**—Amory Lovins**[182]

For decades, Green activists have been attacking our sources of energy. Every single one has been demonized. Coal, which liberated mankind from the Malthusian trap, gave us manufacturing, railroads, and steamships; more than doubled our life expectancy; and saved almost all of us from having to be dirt farmers. Oil, which substantially replaced coal in the 20th century, made airplanes and the private automobile possible, along with the rest of the modern world. And, starting in the 1960s and

'70s, hydropower, nuclear fission, and even natural gas have come under the guns of the activists.

The currently fashionable "renewables," such as wind and solar power, have largely escaped the attacks. Battery-powered electric cars are the darlings of the Greens. But this is because they are simply not capable of providing anywhere near the energy or range that civilization depends on at a price it can afford. Should any of these, or other new forms of energy, prove actually usable on a large scale, they would be attacked just as viciously as fracking for natural gas, which cuts $CO_2$ emissions in half, and nuclear power, which would eliminate them entirely.

Fusion was always going to be the new, clean energy source of the future. As long as it was always 20 years away, nobody minded. But in the early days of the cold fusion phenomenon, it suddenly looked as if the millennium might be at hand, and many of the activists let the mask drop. Their objections really have nothing to do with pollution, or radiation, or risk, or global warming. They are about keeping abundant, cheap energy out of the hands of ordinary people. To a Green, cheap, clean, abundant energy really would be a disaster, because people would gain the ability to change the Earth, and that is, for Green Agonistes, evil.

The classic case, of course, was fission power reactors, the cleanest, safest form of energy ever deployed on a major scale. As a science fiction and technology fan, I had spent most of my life squarely in the "just you wait until we get fusion" camp. Then I was forced to compare the expected advantages that fusion would bring to the ones we already had with fission. Fuel costs are already negligible. The process is already clean, with no emissions. Even though the national waste repository at Yucca Mountain has been blocked by activists since it was designated in 1987, and thus has never opened, fission produces so little waste that all our power plants have operated the entire period by basically sweeping their leftovers into the back closet.

When we compare the excuses that are advanced to suppress fission with its actual record in use, I can assure you that if fusion, cold or hot, ever succeeds, the same excuses, turbocharged and on steroids, will be trotted out against it. In fact, the one significant advantage that fusion might have had was that there was no obvious critical mass for a chain reaction, and thus smaller, cheaper reactors than fission ones might be possible. But the leading fusion project now, the ITER tokamak, is a machine three times the weight of the Eiffel Tower, positioned in a supporting structure the size of the Empire State Building. You will not be using this to power your flying car.

If you are a technologist working on some new, clean, abundant form of energy, I wish you all the luck in the world. But you must not labor under the illusion that, should you succeed, your efforts will be justly rewarded by the gratitude of the people you have lifted from poverty and enabled to have a bright and growing future. You will be attacked, and your invention will be misconstrued and misrepresented by activists, demonized by ignorant journalists, and strangled by regulation.

But only if it works.

# 12

# When Worlds Collide

*Demandez quelque chose, mais le temps.*
(Ask for anything, except time.)

**—Napoleon to his generals**

In the classic science fiction movie *When Worlds Collide*, a group of scientists and engineers gather in a remote compound, engaged in a desperate race to build a rocket ship to escape the imminent destruction of the Earth. Over the entrance is a sign: "Waste Anything but Time!"

The sentiment was familiar to the 1950s audience. It had pervaded the war effort, still the most salient component of culture. The birth of the Atomic Age in the Manhattan Project, for example, had used up all the high-quality graphite in the United States to build the first full-scale reactor, Hanford B, in less than a year. A decade later, at the birth of the Space Age, there wasn't a sign saying "Waste Anything but Time" at the plant where the Apollo spacecraft was being built, but the engineers working there felt like it.[183]

But in the Age of the Eloi, the sense of urgency is gone. Languor and decay have set in. We took a good running start at power too cheap to meter—and fell flat. We had workable flying cars—and we strangled them in red tape. We had Feynman's vision and plan for a technology to rival the Industrial Revolution—and . . . what happened?

The Foresight Institute/*Battelle Roadmap to Productive Nanosystems* was an attempt to chart ways to move nanotechnology research back toward the goals of a broadly based industrial capability at the atomic scale, as originally envisioned by Feynman and Drexler.[184] The word "Productive" in the title was specifically included to indicate this. As a member of the technical working group, I championed Feynman's scheme as one of the possible pathways to molecular manufacturing. It was decided, however, that the *Roadmap* should focus only on techniques that produced atomically precise products. This likely sounds counterintuitive, as if insisting that, at the start of a journey, you grab the map that shows only your destination and excludes your starting point and the intervening territory. And it *was* counterintuitive, except when you factor in the political reasons for it. So much irrelevant technology had been reclassified as "nanotechnology," in a quest both for research funding and cutting-edge status, that without a hard stop in the charter it would have been difficult to accomplish anything.

But the time has long since come to examine the possibility of actually doing what Feynman proposed:

> › Build a machine shop, automated enough that it can be operated by remote control. It has to be capable of building a complete working copy of itself.
> › Also, the shop must be able to do this at its own scale and at a smaller scale. This is, surprisingly, the hard part, because it must fabricate mechanical parts with better tolerances than its own.
> › The goal, of course, is to build smaller and yet smaller copies of our machine shop, but at the same time produce more and more of them. That way, when we get down to machines that can manipulate molecules, we have enough machines to manipulate enough molecules to do something useful.

## Mechanical Motherhood

*Leaping lightly across some centuries of intensive development and discovery, let us consider how the replicator would operate.*

**—Arthur C. Clarke, *Profiles of the Future***

Why has the Feynman Path not been attempted, or even studied and analyzed? One reason is that there still seems to be a "giggle factor" associated with the notion of a compact, macroscale, self-replicating machine using standard fabrication and assembly techniques. Although studied in the abstract since John von Neumann, and in physical systems in biology over roughly the same period, self-replicating machines as a field of engineering remain poorly understood.

One reason for the giggle factor is that we have a strong, well-founded instinct grounded in experience that a factory is much bigger and more complex than whatever it makes. This instinct is neatly captured by a car factory, which is literally thousands of times bigger and more complex than the cars it produces. The Feynman Path toward a self-replicating machine runs counter to that intuition.

George Friedman, professor of engineering at USC and research director of the Space Studies Institute, wrote an extensive analysis of the problems of self-replicating systems:

I can't repeat the many times I've been reassured that self-replication was easy. . . . Biological self-replication can be observed any time we choose to look. [Furthermore], engineering designs can learn from biology, but certain practical simplifications are possible. . . . So if self-replication is so easy, where are all the [self-replicating machines]? [185]

Around the turn of the century, I was in a discussion of the long-term future of nanotech, along with Mark Reed of Yale, a top academic nanotech researcher. I had finished describing the possibilities of atomically precise motors, gears, shafts, pulleys, and the rest, and mentioned that I was quite certain we'd have them sometime in the 21st century. The moderator was incredulous, and asked Reed if he agreed with me. Reed replied (in my certainly not verbatim recollection) that of course we would, but that the machines made from these atomically sized parts wouldn't be able to self-replicate.

One reason for this skepticism, and indeed for the difficulty of designing self-replicating machines (SRMs, or occasionally KSRMs, to distinguish real physical machines from self-replicating software) in the first place, is that they defy standard design methodologies. In standard top-down design, a critical part of the specification for a machine is its capabilities; to design a manufacturing system, for example, it is useful to know what it must manufacture. For a self-replicating system, however, the product is the system itself. You don't know what it has to build until you've finished the design! Therefore, the design process consists of a lot more guessing, cutting, and trying than the normal step-by-step process.

The majority of published SRM designs have either required ready-made complex subsystems (e.g., motors) as the "raw material," resulting in greatly oversimplified construction capabilities, or they have foundered on overly complex capabilities, such as mining and refining naturally occurring raw materials (e.g., the 1980 NASA study aimed at landing an SRM on the Moon and letting it build an industrial base for us). The perception appears to be that there is no middle ground between these extremes.

But things may be shifting. One reason is RepRap, an open-source project to design a 3D printer that you could build in your garage using parts you could get at any hardware store, plus parts that a RepRap could make. RepRap is not a full-fledged self-replicating machine—it can only make about half of the parts it needs, the rest coming from the store. And, of course, making (or buying) parts isn't all the work: You have to put them together. But the project does represent something of a conceptual breakthrough, and a partial answer to the giggle factor. We might say that RepRap is only a quarter of the way to an SRM; however, it is a clear indication that we can "begin to move in that direction."

And RepRap has clearly demonstrated that a 3D printer could be useful in a self-replicating system.

A major milestone toward realizing the Feynman Path would be to work out a scalable architecture for a workable SRM that actually closed the circle. A reasonable start might be a 3D printer, various standard machine-shop items like lathes and milling machines, and a pair of robot arms to move parts from machine to machine and assemble them into new machines. See how close you could get to replication with that, then rinse and repeat.

There are plenty of other problems, but this would at least give us a framework to address them in. *It is a possibility.*

Another reason that the Feynman Path has gone unexplored is the common perception that a machine-based approach had been tried and didn't work.

We have a significant technology of micromachines. The accelerometer and microphone in your smartphone are examples, as is the inkjet head in your printer, and the read/write heads in your hard drive. The labs-on-a-chip that figure in advanced analytical chemistry (and in mRNA vaccine manufacture) are similar.

Today, we make micromachines with the same technology that we use to make electronic chips: photolithography. The Greek roots of the word mean "writing on rock with light." Successive layers of material are deposited on a silicon substrate, exposed like a photographic film, reacted or etched with chemicals, and so forth. When you use these techniques to make a physical machine, instead of (or in addition to) an electronic circuit, the process is called MEMS, for microelectromechanical systems.

It's remarkable how small we can make things this way: machines with actual moving mechanical parts on the scale of a bacterium. But—and this is the key point— they are nowhere near as good as parts that would have been made in a machine shop. We measure the quality of a machine part by relative tolerance: the ratio of the accuracy of its fabrication to its size. In modern machining, relative tolerances of one in a million are standard, whereas in the integrated circuit industry, and MEMS, a relative tolerance of one in 100 is considered good. If I had a steel gear a foot wide, say, as part of a ship's power train, I would expect the tooth faces to have a mirror-smooth finish. But made with the tolerance of MEMS, it's more like you cut the gear out of a chunk of wood with a chainsaw.

The result is that MEMS parts bind, jam, won't slide smoothly, and typically just stick to each other, a sort of super-friction that is called stiction. That means you can't make bearings, gears, screws, cylinders, pistons, ball valves, sliders, or cams that are anywhere nearly as good as they would be in standard machining practice.

A full machining and manipulation capability at the microscale would allow lapping, polishing, and other surface-improvement techniques, which photolithography-based MEMS does not. (Lapping is much like sanding: rubbing things against each other with an abrasive in between makes both of them flatter than before.) You need an appropriately sized machine shop to do those things.

Feynman, in "Plenty of Room at the Bottom," clearly understood both the key importance of precision and the techniques that could be used to obtain it:

> At each stage, it is necessary to improve the precision of the apparatus. If, for instance, having made a small lathe with a pantograph, we find its lead screw irregular—more irregular than the large-scale one—we could lap the lead screw against breakable nuts that you can reverse in the usual way back and forth until this lead screw is, at its scale, as accurate as our original lead screws, at our scale.
>
> We can make flats by rubbing unflat surfaces in triplicates together—in three pairs—and the flats then become flatter than the thing you started with. Thus, it is not impossible to improve precision on a small scale by the correct operations. So, when we build this stuff, it is necessary at each step to improve the

accuracy of the equipment by working for awhile down there, making accurate lead screws, Johansen blocks, and all the other materials which we use in accurate machine work at the higher level. We have to stop at each level and manufacture all the stuff to go to the next level—a very long and very difficult program.[186]

We know, of course, that it is possible to create more precise machines using less precise ones: Our entire industrial base got built in a chain of machines that stretches back to the days when blacksmiths stood under chestnut trees and shaped tools by beating them with hammers. Those of you who are avid amateur astronomers know that you can grind your own telescope mirrors, purely by hand, to a precision much, much tighter than any hand tool can achieve. You do this by an iterative lapping process, much like the one Feynman described.

In other words, the Feynman Path does not envision building a factory that can spit out another factory at quarter-scale and quarter tolerance as if it were stamping out consumer goods. It will require attention and craftsmanship at each level, and probably lots of experimentation and the development of new techniques.

But—to reiterate yet again—at each stage we will have the full fabrication and assembly capability that we need to do experiments, build instruments, and invent and test novel techniques. It is difficult to overstress how valuable this capability, taken for granted at the macroscale, will be at each step to the nanoscale.

## Waste Anything but Time

*To answer the question regarding time span, it seems that a five-year development cycle with that instrument could get you to another plateau of capability; another five-year development from that could get you a long way. I commonly answer that 15 years would not be surprising for major, large-scale applications.*

—**K. Eric Drexler, Senate testimony**

Yes, it would have been wonderful if, back in 1960, people had taken Feynman seriously and really tried the Feynman Path. If only they had, we'd have the full-fledged paraphernalia of real, live molecular machinery now, with everything ranging from countertop replicators to cell-repair machines.

After all, it's been 60 years. The 10 factor-of-four scale reductions to make up the factor-of-a-million scale reduction—from a meter-scale system with centimeter parts to a micron-scale system with 10-nanometer parts—could have been done at a leisurely six years per step. That would have been plenty of time to improve tolerances, do experiments, and invent new techniques.

But now, you object, it's too late. We have major investments and experimentation and development in nanotech of the bottom-up form. We have Drexler's work telling us that the way to molecular manufacturing is not from the top down, starting with machines, but from the bottom up, starting with chemistry and molecular biology.

We have a wide variety of new techniques to read and modify surfaces at the atomic level. We have DNA origami to produce arbitrary patterns and 3D shapes. We have an increasing ability to hijack the existing molecular machinery in the cell and use it to produce the beginnings of usable objects, such as frames and boxes.

Surely by the time a Feynman Path, started now, could get to molecular scale, the existing efforts, including the pathways described in the *Roadmap for Productive Nanosystems*, would have succeeded, leaving us with a relatively useless millimeter-size system with 10-micron-size parts?

No—as the old serials would put it—a thousand times no.

To begin with, a millimeter-size system with 10-micron-size parts would be far from useless. Imagine current-day MEMS but with the capabilities of a full machine shop. The medical applications alone would be staggering.

At this point, a parenthetical observation may be appropriate. The previous paragraph was first written in 2009, as a response to the *Roadmap*. Although I failed to gain enough traction to get any substantial Feynman Path effort going, imagine that I had succeeded. It seems reasonable to estimate that we might have had something like the MEMS-scale machine shop by, say, 2020.

One of the major technological challenges of 2020 was to manufacture mRNA-based COVID-19 vaccines in quantity and at speed. It turns out that the major bottleneck in this manufacturing process is the step in which you produce lipid nanoparticles (LNPs) and introduce mRNA into them. This is done in a microfluidics device that is much like a nebulizer but at a microscopic scale, called an iLiNP. The LNPs come out about 50 nanometers in diameter, but the iLiNP itself is at millimeter scale, with feature sizes ranging from 30 to 100 microns. It's a MEMS device, made with MEMS technology. And there weren't nearly enough of them, nor of the machinery to make them.

You can see where I'm going with this. Even if we had only started a substantive effort along the Feynman Path in 2009, it might well have made an enormous difference in the pandemic.

High-precision machining has not stood still since the '50s. In fact, the starting point today might well be a millimeter-size system with 10-micron-size parts. Existing nano-positioning stages have 50-picometer resolution; existing microgrippers can manipulate parts down to one micron; electron-beam lithography can cut with precision in the 10-nanometer range.

More to the point, a mature technology will need to be able to work at all scales anyway. Imagine that you have a watchmaker's loupe and a set of tweezers that let you precisely manipulate one-millimeter grains of sand. To do the equivalent of making, say, a flying car atom by atom, you would be set the task of building an object the size of the Earth—10 billion parts across. You are going to need a few trucks, at the very least, to augment your tweezers.

If you had followed the Feynman Path to produce your tweezers, they would be robot hands one-sixteenth of an inch wide. And you would *already have* robot arms with quarter-inch hands, and ones with inch-wide hands, and four-inch hands, and hands a foot wide, and five feet, and 20 feet, and 85 feet, 340 feet, 1,365 feet, a mile, four miles, 16, 64, 256—and finally 1,000-mile-wide hands on the end of 10,000-mile-long arms.

The top-down and bottom-up worlds can meet in the middle. When nanoscientists succeed in making an atomically precise nanogear, for example, it means that Feynman Path machines can, once they get to that scale, take the gear off the shelf instead of having to fabricate it. In fact, it is likely that the bottom-up approaches will be the way that parts are made, and the top-down will be the way those parts are put together. I'll stick my neck out and say that if only bottom-up approaches are pursued, we have 20 years to wait for real nanotech; but if the Feynman Path is pursued as well, that time could be cut to a decade.

*It is a long and difficult program—but it is a possibility.*

Taking Feynman's Path to nanotech, or even studying it seriously, would require finding answers to a number of important questions. The answers are invaluable if we are to understand the envelope of possibilities for future manufacturing technology.

Is it possible to build a compact self-replicating machine using macroscopic parts-fabricators and manipulators? We know that a non-compact one is possible—the world industrial infrastructure can replicate itself—and we know that a compact microscopic replicator can work, a bacterium for example. But the bacterium uses diffusive transport, associative recognition of parts by shape-based docking, and other complexity-reducing techniques that are not available at the macroscale.

Not quite the same question: How much cheating can we get away with? In self-replicating machine theory, it's common to specify an environment for the machine to replicate within that contains some "vitamins," or bits that the machine can't make but will need—just as our bodies can't synthesize some of the molecules we need and must get them premade in our diets. For the purpose of the Feynman Path, we can cheat in a very legitimate way: Any part or capability that can be provided on whatever scale it's needed is fine.

We start with control signals—no autonomous AI necessary! If we synthesize molecular gears by chemistry, wonderful. If we can provide single-crystal chunks of silicon or diamond to carve, wonderful. If we can polish surfaces with low-angle electron beams, wonderful. What's left that our machine has to make? What are the roadblocks that we'll have to invent our way around? The classic example is that electromagnetic motors work poorly at small scales, so we have to shift to electrostatic ones.

Near the atomic scale, we hit an "ontological phase boundary," at which we have to quit thinking of our material as continuous (and able to assume arbitrary shapes, like circles) and start treating it as discrete and lumpy. Gravity essentially vanishes, but the adhesion of parts to each other increases tremendously. Lubrication by oily fluids quits working; an oil molecule is an object, not a fluid. Nothing is stiff; even diamond acts like a fairly springy plastic. Hydrogen atoms can jump from point to point by quantum-mechanical tunneling without being anywhere in between. (Luckily, other atoms are too massive to tunnel under most circumstances.)

Given all of this, what kinds of forming techniques will work? Existing 3D printing also can work over a fairly broad range of scales. As far as we know, molding and printing work all the way down until you get to the point where you need complete atomic precision.

Then, when we do need atomic precision, there are a number of 3D-printing technologies we can probably adapt: Instead of melting plastic and squeezing it out like a cake decorator, building a part up layer by layer, we have a supply of molecules that can be made to bond onto the surface by a judicious jolt of current. We can remove molecules in a similar way.

Finally, how do we see what we're doing? It's all very well for Heinlein to talk about scanners, but existing techniques like scanning probes only work over relatively flat surfaces, not in the crowded interior of a milling machine or the like. The answer is that we'll probably have to build physical surface scanners—but existing shop practice makes extensive use of physical contact measurement anyway, so this won't be such a major change.

So we have decided to implement the Feynman Path. What do we do?

## Plan of Attack

*The difficult we do immediately. The impossible takes a little longer.*

**—Motto of the Seabees**

First, for the difficult:

1. Design a scalable, remotely operated manufacturing and manipulation workstation capable of replicating itself anywhere from its own scale to one-quarter relative scale. As noted before, the design is allowed to take advantage of any "vitamins" or other inputs available at the scales they are needed.

2. Implement the architecture at macroscale to test, debug, and verify the design. This would be a physical implementation, probably in plastic or similar materials, at desktop scale, and would include operator controls that would not have to be replicated. Identify phase changes and potential roadblocks in the scaling pathway, and determine scaling steps.

3. Verify scalability of the architecture through these points in simulation. Example: electromagnetic to electrostatic motors. It would be perfectly legitimate to use externally supplied coils above a certain scale if they were available, and below that scale shift to electrostatic actuation, involving only conducting plates, and thus never require the system to be able to wind coils.

4. Write up a detailed, actionable roadmap to the desired fabrication and manipulation techniques at the nanoscale. Note that the state-of-the-art "starting point" (which would be the bootstrap point of the actual scaling series) has nothing to do with the scale (or materials) at which the conceptual debug system, above, would be implemented. The debug system would be at a scale where the parts are easily handled by hand, both to facilitate experimentation and to make the best use of physical intuition in the design process. It would use materials, probably plastics, whose mechanical properties at the macroscale would best simulate those at the nanoscale. The starting-point system, on the other hand, might begin with a particularly stiff material (e.g., electro-discharge machined tungsten carbide) to have as little ground to make up in moving to diamondoid or the like at the nanoscale.

Building a real, physical system is valuable in a number of ways. It is easy to make assumptions in modeling and simulation that the recalcitrant physical world refuses to agree with. Building a working physical model would yield significant insights into capabilities, reveal bugs and design shortcomings, and serve as an experimental platform for improvements.

It would also go a long way toward laying to rest objections about the possibility of SRMs, and serve as a solid experimental data point for further SRM theory.

Thence to the impossible. Identify the best scale, using available fabrication and assembly technology, at which the target architecture can currently be built. It need not be strictly the smallest, if, for example, it appears that we can build a larger one and it can build a smaller one in the time that it would take current technology to build the smaller one.

As a point of reference, for example, at least one experiment has been made that suggests that a one-thousandth-scale system might be achievable. To quote from Robert Freitas and Ralph Merkle's encyclopedic *Kinematic Self-Replicating Machines*:

In 1994 Japanese researchers at Nippondenso Co. Ltd. fabricated a 1/1000th-scale working electric car. As small as a grain of rice, the microcar was a 1/1000-scale replica of the Toyota Motor Corp's first automobile, the 1936 Model AA sedan. The tiny vehicle incorporated 24 assembled parts, including tires, wheels, axles, headlights and taillights, bumpers, a spare tire, and hubcaps carrying the company name inscribed in microscopic letters, all manually assembled using a mechanical micromanipulator of the type generally used for cell handling in biological research. In part because of this hand-crafting, each microcar cost more to build than a full-size modern luxury automobile.[187]

It should be noted that it took a year to make the microcar.

In other words, it's pretty clear that the technology exists to manipulate micron-scale parts, and to make parts with roughnesses of only a few tens of nanometers.

(Roughness isn't the same as tolerance, but in many cases it's just as important.) The Nippondenso microcar had relative tolerances more like MEMS than high-precision machining. However, given the techniques developed since 1994 in mainstream nanotechnology, it's very likely that considerably finer tolerances (and roughnesses) are now possible. If so, we could start the Feynman Path at one-thousandth scale. Another factor of 1,000—just five of Feynman's steps!—and we have nanotech.

A Feynman Path workcell actually avoids the problem that a standard 3D-printer design has with building something its own size, because it's building a copy that's smaller than itself! But, given the techniques that we envision to improve precision, it will have to have general manipulation capability—and the target system, at the nanoscale, will have to build more copies at its own size, so we can't go with just a simple 3D-printer design.

Even so, the key to giving the replicating system a manageably small size is additive manufacturing—3D printing or, more generally, building up a precise part by carefully and repeatedly adding little bits on, as contrasted with standard machining, where we start with a larger chunk of metal and make a precise part by carefully and repeatedly cutting little bits off. It's the main technology that wasn't here in Heinlein's and Feynman's day. The key to using it in a scaling sequence is to understand what kinds of depositions could be done at different stages. One of the most straightforward at the macroscale is melting the substance of interest and allowing it to cool as you deposit it essentially drop by drop on the workpiece. This works for materials ranging from wax to titanium. Scaling works both for and against us here—the superfast dissipation of heat at smaller scales means that you have greater control of what melts in terms of time, but less in space.

At smaller scales, electrodeposition (and electro-removal) will likely have to be used. At the smallest scales, the processes used in electroplating, but controlled at the near-atomic scale, are good candidates. Nanotech enthusiasts have done considerable study of the chemistry of single-molecule deposition reactions in simulation, mostly for diamond synthesis.

A particularly important aspect of the Feynman Path is that, not much more than halfway down to molecular scale in part size, we already hit atomic scale in tolerance. That's within a generation or two from our likely starting point at one-thousandth scale. A micron-sized part really needs atomic-scale tolerance to be considered high-precision. Thus much of the work in that size range will be aimed at surface forming or reforming. Even so, there will be a pressure to design machine elements where bearing surfaces are flat (as in a thrust bearing or slider) so that they can follow crystal planes, until such time as it becomes possible to construct strained-shell circular bearings (simple example: "multi-walled carbon nanotubes").

# Additive Manufacturing

> *I find it incredible that there will not be a sweeping revolution in the methods of building during the next century. The erection of a house-wall, come to think of it, is an astonishingly tedious and complex business; the final result exceedingly unsatisfactory. . . . Better walls than this, and better and less life-wasting ways of making them, are surely possible.*

—**H. G. Wells,** *Anticipations*

If you read the entire passage, Wells is describing, with even better than his usual foresight, the 3D printing of a house.

It's easy to imagine the kinds of improvements that additive manufacturing will see over the next decade or two. The speed of big commercial machines will increase, and the cost of personal ones will decrease. The range of usable materials will increase. Precision will get better. We will get closer and closer to being able to print out anything that today is factory manufactured—"Whatever you want, whenever you want." And let us add "Wherever you want." This means not just plastic, or even metal shapes, but working machines. Ultimately, even working 3D printers.

Although classic SF writers didn't quite seize on nanotech as the way that replicators would work, they did realize that personal-size replication could have major economic, and thus social, consequences. The best-known example of this was in George O. Smith's *Venus Equilateral* stories; the result is the breakdown of civilized society. That made for a rollicking story, but in the real world pretty nearly the opposite happened. Cars overtook trains. Cellphones overtook landlines. One-hour photo shops overtook the practice of sending your film to Rochester to be developed. Digital cameras overtook one-hour photo shops. PCs overtook the horridly bureaucratic corporate (and university) computer centers, and reduced the cost of computing by a factor of a billion. In no case did the world fall apart.

I predict that the nanomanufacturing revolution will resemble the computer revolution in many ways. And I imagine that it will have many of the same kinds of salutary effects, as far as personal autonomy and the enhancement of creativity. I would say that in the 1970s there were on the order of 100 software applications available to computer users in the big computer centers where I did my work as a student; today there are 1.5 million apps for your smartphone. It's not just a case of anything you can imagine at your fingertips; there are so many things available that you would never have thought of by yourself. Imagine the same "Cambrian explosion" with physical machines.

This would be, in a strong sense, a resurgence of what happened in the Industrial Revolution.

Norwegian archaeologists recently uncovered a woolen tunic that had been abandoned on a trail across the Lendbreen glacier 1,700 years ago.[188] They were surprised to find that something so valuable had apparently been abandoned. In order to give their audience a sense of how valuable it would have been to its original owner, the archaeologists made a film in which they re-created the process that people of the year 300 would have used to make it. This involves hand-picking the wool from sheep who were shedding their coats, and hand-spinning thread, and weaving this thread using a primitive loom, followed by cutting and sewing and so forth, all with simple, unpowered hand tools.

Adam Smith, founder of economics, found this sort of process remarkable:

> The woollen coat, for example, which covers the day-labourer, as coarse and rough as it may appear, is the produce of the joint labour of a great multitude of workmen. The shepherd, the sorter of the wool, the wool-comber or carder, the dyer, the scribbler, the spinner, the weaver, the fuller, the dresser, with many others, must all join their different arts in order to complete even this homely production.[189]

By Smith's day, patterns of specialization and trade had created something that we would call a supply chain; but this description underscores just how much work had to go into the tunic. Ironically, Marianne Vedeler and Lena Hammarlund, the archaeologists, quickly discovered that it would be too expensive to re-create an entire tunic the old way, and were forced to use machines to complete their re-creation.

The total amount of labor necessary to create one tunic worked out to be three months of full-time work. At the average wage in the United States, as of 2020, that would make the tunic cost $15,000. By contrast, one can go on Amazon today and get an equivalent woolen tunic for $50, or two and a half hours' work. The difference, of course, is the Industrial Revolution. Because of the machines that we've invented, each hour of human labor today produces 300 times as much as it did seven centuries ago. Your medieval counterpart would spend a year to make as much as you make in a day.

The promise of nanotech is that this could happen again. Things that now take us a year's work could be done in a day. And your $3,000,000 flying car would cost just $10,000.

*It is a possibility.*

# 13

# When the Sleeper Wakes

*For ages man lived in a world where he was a slave to the elements. His own achievements were by comparison crude and immature; his every living moment was subject to the blind caprices of fate. Not unnaturally, he dreamt of greater things. It was not until man found himself capable of transforming dreams into prophesy that he wrote science fiction. The only difference between the science fiction fan of today and the Homer of yesteryear is that the fan of today knows that there is a sufficiently large kernel of truth in his dreams to make them possible of realization— that the fantastic fiction of today may well become the fact of tomorrow.*

—**Sam Moskowitz,** ***The Immortal Storm***

The question of flying cars has long been one flung back at technological futurists. Around the turn of the century, some of the researchers in the NASA Ames nanotech effort asked me to look at what the capabilities of mature nanotech could do to build flying cars. I found the technical problems interesting but not difficult. Becoming a pilot and airplane owner in the meantime has changed my appreciation of some of the details, but not the main conclusion. But there the question rested for some time.

There now appears to be some renewed interest in "What happened to the future we were promised?" Perhaps it is partly due to the 50th anniversary of the Space Age. Perhaps it is partly due to the fact that we can see how far computers have progressed, and wonder why other things have not progressed nearly as much. Perhaps we look back on the optimism of our grandparents for the future, and wonder what happened to it. But no matter the reason, the general impression still seems to be that flying cars, and all the other hoped-for technologies, were over-promised.

This gets it almost completely backward. Flying cars of the convertible type are technologically straightforward. Not only are they easily within reach of the current capabilities, but they have been built, and successfully flown, with almost monotonous regularity by individuals and small companies since the 1930s. The autogyro had both the innovative brilliance that would have allowed short home runways and a dedicated, experienced, and well-funded sponsor.

Unfortunately, in the 1930s, marketing a new, non-mass-produced vehicle to Depression America was not a winning proposition.

In the 1940s, World War II sucked virtually all our engineering talent into war production. The size of the aviation industry quadrupled. The average American was just as unable to afford a car as he had been in the depths of the Great Depression. In the postwar '40s, several quite workable convertible airplanes were developed.

In the 1950s, delays associated with aircraft regulation blunted the supply side of flying cars, while the burgeoning investment in superhighways, bridges, and other infrastructure blunted the demand side. The government spent an inexplicable amount of time and resources trying to prevent Pitcairn from getting his due, incidentally preventing him from spending his time, energy, and resources on developing personal flying machines in the postwar era. Even so, a handful of convertible flying-car types got aircraft certification and came close to production. All in a world where real disposable income was a third of today's. Helicopters developed into a practical form of transport but remained too expensive for the average family.

In the 1960s, passenger jet travel came into its own, becoming a major new mode of transportation. Alongside that, the '60s were a heyday of private aviation; a reasonable extrapolation of the trend would have put a fair number of us in flying cars today. In the 1960s, the military developed perfectly workable (if noisy) car-size VTOLs (e.g., the Airgeeps).

In the 1970s, the centuries-long growth trend in energy (the Henry Adams Curve) flatlined. Most of the techno-predictions from the science fiction of the 1950s and '60s had assumed, at least implicitly, that it would continue. The failed predictions strongly correlate to dependence on plentiful energy. American investment and innovation in transportation languished; no new developments of comparable impact have succeeded highways and airliners.

Also in the 1970s, academia became a major locus of countercultural fervor, as it morphed into a virtue-signaling institution driven by competitive self-deception. This set up a classic Baptists and Bootleggers bedfellowship between those who really believed that progressive prescriptions would improve the world and those who mostly enjoyed the new money, prestige, and policy-making influence. Public spending, and Ph.D.'s granted, tripled between 1960 and 1980. In the 1970s, the war on cars was handed off from beatniks to bureaucrats. Supersonic flight was banned. Bridge building had peaked in the 1960s, and traffic congestion now is five times as bad as it was during that decade.

Around 1980, developments in liability law destroyed the private aviation industry. Regulation exploded; a significant proportion of decisions in business went from being made by people who were forced to balance costs with benefits to being made by bureaucrats with no concern for costs. In our economy, the cost disease increasingly replaced the learning curve. The nuclear industry found its costs jacked up by an order of magnitude, and was essentially frozen in place. Interest and research in nuclear physics languished.

Over the period of stagnation, Green fundamentalism became the unofficial state church of the U.S. (and, to an even greater extent, Western Europe). Its catechism is a litany of apocalyptic prophecies, each one's details forgotten as it failed, but each adding to an overall angst of original sin and impending doom. This has contributed in no small part to the neurotic pessimism of our current culture, by objective measures the richest, safest, and healthiest in history.

In technological terms, the bottom line is simple: We could very easily have flying cars today. Indeed, we could have had them in 1950, but for the Depression and World War II. The proximate reason we don't have them now is the Henry Adams Curve flatline; the reasons for the flatline have taken much of this book to explore. We have let complacent naysayers metamorphose from the pundits of a century ago muttering "It can't be done" into today's bureaucrats thundering "It won't be done."

The golden-age science fiction writers understood the technology for things like flying cars reasonably well. They understood the potential of nuclear power fairly well. The one thing that they really over-promised was fusion—they were about 50 years optimistic, at a guess. And even then their problem arose from listening too closely to the accredited experts of the day.

Ironically, the reasons for mispredictions were opposite for flying cars and for space travel. In the former case, a technology that would have developed normally over the century was hobbled by the Great Depression and World War II, hounded by a nearly inexplicable postwar government-led legal effort, stifled by regulation, and then essentially extinguished by liability lawyers.

On the other hand, rockets and space flight were highly funded by governments, for military reasons and as programs for national prestige. By the '60s, some writers (notably Clarke) were beginning to mistake that for sustainable progress, producing an overly optimistic view of things like planetary colonization in our lifetimes. Heinlein was much closer to the mark. He predicted a takeover of the U.S. government and ruling class by a religion hostile to technology, and a hiatus in space travel.

The abortive Atomic Age, with its unkept promise of energy too cheap to meter, combined elements of both the flying car and space travel in its demise. During World War II, the government spent an enormous amount of money and effort developing nuclear weapons. This gave pundits, including science fiction writers, reasons to believe that we had solved the mysteries of the atom and had a great new energy source available to us. In fact, the huge investment in the atomic bomb—remember, the Manhattan Project built more industrial plants than the entire American automotive industry—was a dead loss economically.

The Manhattan Project cost roughly $2 billion in nominal dollars. The only single project that cost more was the development of the B-29 fleet, which cost $3 billion. The investment and technological experience in the B-29 did pay off economically. The postwar years benefited from large, pressurized, high-altitude airliners, and their continued development was commercially sustainable. By contrast, the military restrictions on anything nuclear hobbled private experimentation and development. It took nearly two decades from the first supersecret "atomic piles" to an economically useful nuclear power sector. Then, within another couple of decades, regulation petrified the industry.

In contrast, during the 1970s and '80s, computing went through a phase change, from bureaucratic computing centers in corporations and universities to a multitude of smaller, private, unregulated machines. Combined with the growth of the internet in the '80s and '90s, this kept information technology essentially on the track that the technological optimists predicted. Asimov and Clarke got the state of applications, ranging from robots to planetary libraries, more or less right. The experience of ordering something online (i.e., with video pictures on a home console) and having it delivered within a day is flabbergastingly close to what Philip Francis Nowlan predicted in 1927 and Heinlein gave a more detailed picture of in 1938. The main difference is that physical delivery takes longer.

The rest of technology, particularly anything high-energy, remains in the computer-center era. Airline travel is Exhibit A.

## Science

In the background, scientific knowledge has kept advancing, Great Stagnation notwithstanding. The rate of technological progress in areas without opposition or strangulation, such as computer technology, has been tremendous. During the decades of stagnation, basic science has discovered planets around other stars, and indeed increased the number of known bodies in our own solar system by a factor of 1,000. We have solved the molecular mechanisms of life itself, both working out the molecular shape and function of the genome and then reading it; we have also mapped the human brain, producing remarkably detailed wiring diagrams. We have built giant underground neutrino telescopes and measured the 450 billion neutrinos per square inch per second that sleet through our bodies direct from the nuclear processes at the center of the Sun. We've detected a flash of gravitational radiation from a stellar system collapse so violent that three solar masses of matter were converted into energy in the form of waves in the fabric of space-time itself.

A substantial number of the expected but unattained technological advances were not mispredictions of technology. They were instead misplaced faith in the vitality and dynamism of our culture, and in the wisdom and, indeed, the basic competence of our information providers and systems of governance. We poured increasing torrents of money into the ivory tower and virtue signaling, and it increasingly took our best and our brightest away from improving our lives.

When, in 1954, Allen Ginsberg wrote, "I saw the best minds of my generation destroyed by madness," he was projecting, but by 1974 he was not half wrong. Imagine that the best and brightest of my classmates had spent their time studying instead of protesting, working out new inventions together instead of getting stoned together, and had gone on to become engineers instead of activists, regulators, and lawyers. Imagine that education, health care, and public works had not been overcome by the cost disease. Imagine that the private aircraft industry had grown instead of being destroyed; that the Henry Adams Curve had not flatlined; that technology had continued to develop, and life to improve, the way that it had done in the previous 50 years.

Purely from a technological standpoint, what would life be like now for the typical American?

To begin with, the average American would be about four times as wealthy; call it a median $200,000 family income, as opposed to today's $50,000. Approximately the top 5 percent of families now can afford two homes; without the Stagnation, the top 50 percent would be in the same bracket. This is essentially the demographic that could, if they wished, own a helicopter, or a 200-mile-per-hour multi-fan tilt-rotor flying car.

The makeup of the "products" that constitute the GDP would be quite different. It would run more to actual machines and less to expensive paper-pushing. (The size of the financial sector has doubled as a fraction of America's GDP since 1980; it's bigger than the energy sector. More dollars, less value. Both the educational and medical sectors today rake in 10 times as much, in relative terms, as they do in other countries, while producing similar results.) There would have been more experience with things high-power and high-tech, and so we would be much further along the learning curve with, for example, turbine aero engines.

The April 4, 2015, edition of *The Economist* featured as a cover story a study by the London School of Economics capturing the economic history of cities. It noted:

> In the 20th century, tumbling transport costs weakened the gravitational pull of the city; in the 21st, the digital revolution has restored it. Knowledge-intensive industries such as technology and finance thrive on the clustering of workers who share ideas and expertise. The economies and populations of metropolises like London, New York and San Francisco have rebounded as a result.[190]

They somehow missed the fact that *transport costs quit tumbling.* I doubt that there is a more comprehensive nutshell for the Great Strangulation. A major component of the opportunity cost of travel is the time taken. As the highways became more congested, time cost rose; traffic congestion today is five times worse than it was 25 years ago. But planners (including the study authors!) have increasingly adopted a car-hostile ideology, and are designing cities intentionally to make driving more costly and difficult.

If transportation costs had, in fact, kept tumbling, particularly in terms of time, we would have seen a continuation of the great push out into the exurbs. The difference is that a flight-based regime of expansion not only expands the sphere of influence of an urban center, it makes multiple urban centers *and all the space in between* available to the rural dweller. And that, in turn, means that it becomes viable to put a business just about anywhere.

A big watershed would come when it became reasonably economical to build and occupy a mountainside home without building a road or physical utility connections. UPS drops by every day, in a big brown helicopter, and your own family VTOL takes care of most simple transportation needs. Telecom is easy—we've essentially got that today. But what about power?

Cover the roof with photovoltaics for solar power backed with batteries, or get it out in the form of fuel that you then feed into fuel cells. As I write, both conversion steps are seeing hints of happening in laboratories, and would probably just be becoming practical with a non-stagnated nanotech. Unfortunately, this doesn't provide enough energy for a Level 5 lifestyle. Periodic deliveries of chemical fuel wouldn't either. Given a vigorous, non-stagnated nuclear power technology, several generations on from our current dinosaurs, something like NASA's 10-kilowatt Kilopower reactor is at least conceivable. As I write, the development of the Kilopower is being superseded by a 20-kW half-ton "Chargeable Atomic Battery," which is claimed to be "easier and cheaper to manufacture than Pu-238" (i.e., for RTGs), and to have 30 times the power density.[191] This is getting fairly close to isotopic batteries (direct electric, not thermal), as foreseen by Asimov. If you don't trust Asimov's (and my) technological intuition that 50 years' research and innovation in the field would have made direct conversion straightforward by now, note that the technology of thermal conversion hasn't stood still since the Eisenhower administration either.

If, instead, we had built a cellular power system with small modular reactors of a more or less conventional design every few miles, we could transmit power to our mountainside house. Transmitted power was studied in detail in the 1970s, as part of Gerry O'Neill's High Frontier effort, which proposed building space colonies and solar power satellites. Our electronics have advanced considerably since then; so has our rocketry. The whole plan should be resurrected, and the remote home could get power directly from the sky.

On the other hand, not only do we live beneath an enormous natural fusion reactor, we live on top of an enormous natural fission reactor. At 6,000° Celsius, the center of the Earth is, somewhat surprisingly, as hot as the surface of the Sun. Its temperature is maintained by the ongoing fission of uranium. Throughout much of the world, we need only drill down about 10 miles to tap into essentially inexhaustible energy. The ability to do that economically is on the edge of feasible. Had technology not stagnated, that might today be a growth industry.

The technological possibilities are endless; it's a question of what's more economical.

In any case, we can confidently guess that, without stagnation, the diaspora from the cities would have continued rather than reversing. Destinations would include the sea, the mountains, and other areas on the Earth's surface. There would be substantially more space travel and exploration, but not as yet substantial colonization—there's an enormous amount of vastly underutilized room here on Earth. A major push to live in space will not happen until we have mature, full-fledged nanotech. In a pro-technology alternative present, we would be seeing the earliest real nanotech applications right about now, much like personal computing in the 1980s. And early adopters would be just starting to switch from turbine/generators to fuel cells in their flying cars.

Today, there seems to be a Cambrian explosion in proposed flying-car types, and many of them are battery/electric. There is probably more money than is strictly wise being invested in eVTOL taxis. At my best reading of the technical literature, the headroom for batteries in known chemistries is about a factor of eight in energy density, while perhaps cutting the price in half. This is certainly good enough for ground cars and even fixed-wing airplanes, but not VTOLs, which, according to travel theory, need a 500-mile range to attain their best value. On the other hand, fuel cells appear to be fairly low-hanging fruit for early nanotech, and they could be a drop-in replacement for aircraft that had been developed using batteries. So I am inclined to view the current electric-flying-car enthusiasm as being more sustainable than not.

In fact, if you look carefully, it is even possible that we could be on the verge of a technological renaissance of a sort. About 10 years ago, I had pointed out that anhydrous ammonia might make a useful fuel for fuel cells, being easier to store and handle than hydrogen yet equally carbon-free. A decade on, there is a substantial body of literature, conferences, and companies at the point of viable commercialization.

There are startups exploring small modular reactors, both fission and fusion. Biotechnology has taken off in a big way. We have drones, videophones, digital assistants, and robots that do a half-decent job of vacuuming. And we are traveling into orbit in reusable rockets that land themselves on tails of fire, as God and Robert Heinlein intended.

There are many new technologies that may appear in the coming century, some more anticipated than others. Will some of them manage to re-create the miracle of the Industrial Revolution, and improve the life of the ordinary person by an order of magnitude? We shall see.

# III

# Profiles
of the
Future

# 14

# The Dawn
# of Robots

*And now for the World of the Airmen and a new start for mankind.*

—H. G. Wells, *Things to Come*

After the flying car, the thing that people seem to remember most about *The Jetsons* is Rosie the robot housemaid. Robots, it turns out, are a harder problem technologically than flying cars. Rosie didn't get stagnated; the state of robots and artificial intelligence now is pretty much just as Asimov predicted half a century ago. As we pivot to look at the future, we must try to do as good a job. We shall plant our feet on as firm a ground of current capabilities in each area as we can, and then look forward, attempting to avoid Failures of Nerve (not hard) and of Imagination (not easy).

A decade ago, in *Beyond AI*, I took up cudgels against those who were predicting a major takeoff in artificial intelligence by virtue of a self-improving super-AI. Long before that happens, I said, we will see something a bit more mundane but perfectly effective: AI will start to work, and people will realize it, and lots of money, talent, and resources will pour into the field.

> It might affect AI like the Wright brothers' Paris demonstrations of their flying machine did a century ago. After ignoring their successful first flight for years, the scientific community finally acknowledged it; and aviation went from a screwball hobby to the rage of the age, and kept that cachet for decades. In particular, the amount of development effort took off enormously.[192]

The amount of money going into aviation before 1910 was essentially nil (Langley's grants notwithstanding). Once people caught on that airplanes really worked, though, there was a sensation and a boom. Accelerating results attract more money—there's your feedback loop. By the end of the 1920s, Pan American was flying scheduled international flights in the eight-passenger Ford Tri-Motor. The ensuing exponential growth in capabilities, which continued unabated right up to the '60s, was very much part of the zeitgeist of the "future we were promised."

Something of the kind appears to be happening in AI as I write. In just the past few years, a breakthrough from an unexpected source, deep learning in multilevel neural networks, has brought significant new capabilities. In some sense, "deep learning" is just a buzzword phrase, since the technique is basically the same back-propagation of error terms that has been used since the 1980s. But like the advances in self-driving cars, a lot of small improvements have combined with substantial new processing power, in the form of graphic processing units (GPUs). Raw computing power lets us train networks with 22 levels, whereas the standard a couple of decades ago was two or three. We can train a neural net in an hour that in the 1990s took weeks. And the availability of large datasets for training has left us with the ability to give the network a much more comprehensive, broad-based experience and background in something approaching (at least verbal) common sense.

These, of course, are combined with (and enabled by) the progress in various other parts of the field of AI, ranging from processing power to robotics. A decade ago, we did not have robots who could learn manual tasks by watching them being done and practicing them; now we do.[193] A decade ago, you were better off fat-fingering your text message into your smartphone and suffering the indignities of autocorrect than you were trying to use speech recognition. In my recent experience, that has changed: Now speech recognition works better than typing.

As an experiment to see what the state of the art is like, I took a popular open-source collection of learning tools and trained a multilevel recurrent neural net on a 50-megabyte corpus of adventure and science fiction novels.[194] Untrained, the model produces complete random-character gibberish, like "w98vWma0N&w47t*Knq4." After one night's training on my not particularly powerful workstation, it was capable of producing something much more English-like:

> The jets of the hundred for the idea of the Minypour from a seemed planet, now on when the left the band should blow that to have him from glarge. But they were always for the grim, though even for the very arm-visions.

And after another couple of days, you had to pay attention to realize it was nonsense:

> "Dejah Thoris," he said. "I'll be somewhere, which may be the meaning of all the cabin. But I know that you've got some of his heart all your teleceiver of the Lensman."

The neural net had started with no knowledge whatsoever, even implicit, of the structure of language. By the time I took that second sample, it had picked up the idea of words, accumulated a good vocabulary, understood enough of sentences to capital-ize and punctuate, and enough grammar and parts of speech to produce coherent prepositional phrases. It fell a bit short of full subject-predicate sentences after one night, but a couple of days later was getting close. When the net did decide to make up a word, it produced something reasonable, pronounceable, even creative.

The reason this is amazing is that the task the model was set to learn was simply to predict the next character, given a sliding 50-character window in the text. It worked out all of the concepts and structure completely from scratch. This is an ex-tremely powerful capability. It may very well take another decade to integrate it into the full panoply of AI techniques that have been developed over the past 50 years, from symbolic logic to intelligent agents to evolutionary algorithms. But it "seems fairly safe prophesying" to guess that by the end of the 2020s we will have something that has at least humanlike performance at any task involving reading, writing, talking, and listening.

In 2015, the Google company DeepMind brought out AlphaGo. AlphaGo was the first program capable of playing the game Go at a professional level. A mere decade ago, as I described the situation in *Beyond AI*:

> By contrast to chess, programs playing Go have yet to challenge even serious amateurs, much less the top professionals. . . . The connected clusters on the Go board are like the pieces in chess. But there are only six different kinds of chess piece; the number of different configurations of a cluster on a Go board is astronomical. So in playing Go, if you equate clusters to pieces, you have to be able to invent new pieces as you go, so to speak, and to infer the capabilities of the ones your opponent is inventing. Thus, at the moment, i.e. unless someone finds another cute trick that is a special case solution, playing Go well requires solving the real AI problem, that of creating useful concepts from experience.[195]

In other words, playing Go well requires the same kind of concept formation that was demonstrated by the deep neural net in climbing from character sequences to words, phrases, and sentences. You will not be at all surprised that AlphaGo contains not one but two deep neural nets, which are effectively integrated with a classical algorithmic game-playing program.[196]

AlphaGo improved with shocking rapidity, beating Lee Sedol, a top-ranked Go player, in a million-dollar challenge tournament in March 2016.

At the time of the tournament, virtually every commentator understood the value of the technical innovation behind the working combination of algorithmic search and learning neural nets. Many observers, however, seemed to share the same ivory-tower blindness that my friend had showed toward the DARPA self-driving cars challenge. AlphaGo is a scalable program, running on a cluster of up to 1,920 central processing units (CPUs) and 280 GPUs. That allowed some people to denigrate the feat as "brute force," and thus not impressive from an algorithmic perspective.

But in the real world, what is impressive is that AI researchers now have the resources to use whatever horsepower is necessary, and the know-how to deploy it effectively. In fact, it was revealed a little later that Google had built custom chips specifically to accelerate the deep-learning algorithms, which had given them about a half-decade jump on expected Moore's Law progress.

Another interesting indication of the fact that AI is moving from academia to industry is that DeepMind's *Nature* paper about AlphaGo has 20 authors. And LinkedIn's 2017 Jobs Report told us that the fastest-growing position is machine learning engineer. And there were at least 225 AI/machine-learning conferences held in 2018. Perhaps the brothers Wright are gazing down with an amused sense of déjà vu. We live in interesting times.

## Robo Habilis

*A human being should be able to change a diaper, plan an invasion, butcher a hog, conn a ship, design a building, write a sonnet, balance accounts, build a wall, set a bone, comfort the dying, take orders, give orders, cooperate, act alone, solve equations, analyze a new problem, pitch manure, program a computer, cook a tasty meal, fight efficiently, die gallantly. Specialization is for insects.*

—**Heinlein,** *Time Enough for Love*

In an interview a few years ago, Steve Wozniak of Apple fame opined that there would never be a robot that could walk into an unfamiliar house and make a cup of coffee. I feel that the task is demanding enough to stand as a pons asinorum for an embodied AI, that is, a robot in the sense that Asimov used the word.

A robot is placed at the door of a typical house or apartment. It must find a doorbell or knocker, or simply knock on the door. When the door is answered, it must explain itself to the householder and enter once it has been invited in. (We will assume that the householder has agreed to allow the test in her house, but is otherwise completely unconnected with the team doing the experiment, and indeed has no special knowledge of AI or robotics at all.) On entering, the robot must find the kitchen, locate coffee-making supplies and equipment, make coffee to the householder's taste, and serve it in some other room. The robot is allowed, indeed required by the specifics of the test, to ask questions of the householder, but it may not be physically assisted.

The current state of the robotics art falls short of this capability in a number of ways. The robot will need to use vision to navigate, identify objects, possibly identify gestures ("the coffee's in that cabinet over there"), and to coordinate complex manipulations. Manipulation and physical modeling in a tight feedback learning loop may be necessary—for example, in order to pour coffee from an unfamiliar pot into an unfamiliar cup. Speech recognition and natural language understanding and generation will be necessary. Planning must be done at a host of levels, ranging from manipulator paths to coffee-brewing sequences.

But the major advance for a coffee-making robot is that all of these capabilities must be coordinated and used appropriately and coherently in aid of the overall goal. The usual narrow AI formulations of these problems—setup, task definition, and so forth—are gone; the robot has to find the problems as well as solve them. That makes coffee-making a strenuous test of a system's adaptability and common sense.

I claim that this test addresses the bulk of the aspects of general intelligence that are missing from AI today. Although standard shortcuts might be used, such as having a database of every manufactured coffeemaker ever built, it would be prohibitive to have the actual manipulation sequences for each one preprogrammed, especially given the variability in workspace geometry, different containers of coffee grounds, and so forth. Transfer learning, generalization, reasoning by analogy, and, in particular, learning from example and practice are almost certain to be necessary for the system to be practical.

Coffee-making is a task that most 10-year-old humans can do reliably with a modicum of experience. I would guess that a week's worth of example and practice in a variety of homes would be enough for a 10-year-old to pass such a "Wozniak Test" in a majority of homes. It is a good test of generality in a robot because, although it would be possible to hand-code most of the skills needed, it would be much cheaper simply to build a coffeemaker into the robot. Consequently, nobody would spend the resources necessary for the brute-force programming approach. The only economical way to do the task is to build general learning skills and have a robot that is capable of learning not only to make coffee but to do any similar domestic chore.

One of the species of early hominids is named *Homo habilis*, meaning "handy man," for their significant advance in tool use over previous hominids. One of the goals of the AGI Roadmap was to chart a path from the current state of AI to full human intelligence.[197] The Wozniak Test, as a test not only of tool use but tool recognition, would be a major milestone on that path. It turns out to be a special case of what we might call the Nilsson Test, as outlined in a 2005 paper by Nils Nilsson, one of the leading figures in AI:

> Machines exhibiting true human-level intelligence should be able to do many of the things humans are able to do. Among these activities are the tasks or "jobs" at which people are employed. I suggest we replace the Turing test by something I will call the "employment test." To pass the employment test, AI programs must be able to perform the jobs ordinarily performed by humans.[198]

The forward step in AI represented by AlphaGo does not imply a comparable step in physical robotics. One of the main components of the surge in efficacy of deep-learning neural nets was the availability of enormous amounts of data on which to train them. For language machines, there were literally millions of books and other textual data and audio available online; for AlphaGo, the training corpus included the historical record of human games played. (The earliest recorded Go game was in AD 250, but records only go back about 500 years.) In addition, AlphaGo played millions of games against itself in the process of learning.

This works much better in the simple, well-defined, digital environment of a board game—a typical game of Go can be compressed into about 237 bytes,[199] and a chess game into less than 100—than in the boisterous, messy, expensive real world. Ironically, in the same week that Google's DeepMind company won with AlphaGo, they put their robotics company, Boston Dynamics, up for sale. Gary Marcus, CEO of the AI startup Geometric Intelligence, quipped, "Rosie the robot, you can't have it knock over your furniture a hundred thousand times to learn."[200]

This points to one of the major weaknesses of the modern neural-net-based techniques of AI. While remarkably good at learning (much better than our other attempts), they are nowhere near as good as humans at learning fast, from limited experience. This is, of course, something that AI researchers know very well and are putting a lot of work into. After that will come learning by imitating, learning by being told how, and learning by trial and error.

The bottom line is that today we see only the first light of the dawn of the Age of Robots. But we can see enough by this light that failing to expect these machines has changed from a Failure of the Imagination to a Failure of Nerve. However, even that failure is not common anymore; of all the future technologies I have discussed, robots are closest to happening on schedule, and the general run of expectations is the most realistic.

## Profession

> *Those were terrible days for the Americans. They were hunted like wild beasts. Only those survived who finally found refuge in mountains, canyons and forests. Government was at an end among them. Anarchy prevailed for several generations. Most would have been eager to submit to the Hans, even if it meant slavery. But the Hans did not want them, for they themselves had marvelous machinery and scientific process by which all difficult labor was accomplished.*

**—Philip Francis Nowlan, *Armageddon 2419 A.D.***

*The Jetsons* was a caricature of 1960s-era life, and thus left unexplored many of the implications of the trends it depicted. For example: Rosie was easily smart enough to do everything that George Jetson did at work. So, why didn't Spacely Space Sprockets simply buy Rosie instead of hiring George?

In real life, of course, this is essentially what has happened. We haven't replaced all our workers with robots, but in the manufacturing sector we've replaced 85 percent of them. This only continues a trend. Since the Industrial Revolution, advancing technology has been making it possible for fewer and fewer people to produce more and more stuff.

Back when George Washington was president, virtually everybody in America (and everywhere else in the world) had to work on a farm just to produce enough food to live. Each farmer only produced enough food for 1.1 people. But as technology improved, farmers became more efficient, and each farmer now feeds 40 people. There does, of course, need to be industry to produce the machines that modern farmers use. But the same trend can be seen in manufacturing. In the 1940s, over a third of jobs in the U.S. were in manufacturing; now less than 9 percent are.

The number of manufacturing jobs per capita has declined steadily since World War II, just as the farm jobs have. This is not because industry is producing less, or because jobs have been going overseas, as is sometimes claimed. After all, domestic

American factories produce three times as much in total goods as they did in 1950. Output might have risen faster without offshoring, but it certainly did not decline.

But there's another way of looking at the manufacturing statistics, much the way we look at how each farmer feeds 40 times as many people as before. The number of manufacturing workers went down and the output went up—and the output per worker went way up. Each manufacturing worker in 1962 produced about $25,000 (in constant 2010 dollars) of output; in 2012, each worker produced $160,000.[201] It may help to remember that this includes not only the assembly-line workers but also foremen, engineers, quality-assurance specialists, accountants, managers, and lawyers—everybody who works for a manufacturing company. Productivity is clearly on an exponential increase; in fact, if you fit a curve to the data, you get a curve that is slightly hyperexponential.[202]

If things keep up at this rate, by 2062 the average worker in manufacturing (George Jetson, for example) would produce $661,000 in value per year. Note, for comparison, that the average employee of Apple Computer in 2012 already produced $400,000 in value; so did the average Lockheed Martin employee. Over the latter 20th century, workers have gotten paid about 64 percent of the value added, with the rest going to taxes, capital equipment, dividends for stockholders, and so forth. At this rate, George Jetson makes $423,000 in 2062. (Although, if we stay with the same roughly 3 percent inflation rate we've averaged since the 1980s, he will call it $1,854,392.)

The hyperexponential nature of the fit, though slight, is significant. In a pure exponential growth curve, the year-to-year percentage increase is fixed. With the hyperexponential curve, it increases. In 1962, a manufacturing worker was 1.1 percent more productive each year. By 2062, Jetson would be 3.7 percent more productive each year. Add this to the cumulative nature of the curve and the result is startling: In 1962, the average worker produced $734 more than in 1961. In 2062, Jetson produces $23,562 more than in 2061.

But will he work at all? Let us do a thought experiment.

Interstate highway I-80 runs just about 3,000 miles, from New York City to San Francisco. As of 2021, the population of the United States is (very roughly) 300 million people. Line the people up along I-80; give them a square yard apiece, and they will cover the whole 180-foot-wide right-of-way with 100,000 people per mile. Now, what we are going to do is to line them up in order of economic productivity—assuming that they work full-time—the most productive standing in New York, the least in San Francisco.

The point of the exercise is to find the point where, with a given technology, you would have to separate workers from nonworkers if you wanted to have the fewest people working in the given economy. Obviously, you would employ the most productive, since that way you get the most output from the fewest people. You employ everyone to the east of your chosen point. Also, obviously, that point would depend on the total amount of consumption you wish to have in the society.

In 1790s America, virtually everybody had to work to maintain just a level of consumption at approximately Rosling's Level 2—people had shoes but no bicycles. The same was true of all societies in the world up to that time. If 90 percent of workers were farmers, there weren't too many people left to be butchers, bakers, and candlestick makers. This means that everybody standing along I-80 outside of California had to work.

So here's an interesting question: Where would we put the dividing point along the highway if we substituted the technology of today, but kept the consumption levels of 1790? We've noted that each farmer today feeds 40 people instead of 1.1, but a surprising number of today's farmers either don't provide subsistence-level food or don't use the best technology. A real-world example, not far from my house, is an organic lavender farm. What is more, Americans today eat a lot more food, and a lot more varied food, than our ancestors did in 1790.

Thus, I feel safe in guessing that, given 2020s technology but requiring merely a 1790s level of consumption, only the people standing in New Jersey would have to work. In other words, an industrial cadre of just six million (with whatever capital they needed) could feed, clothe, and shelter the rest of the country as well as the average American had been fed, clothed, and sheltered during the Washington administration. (Do remember that the six million are, by assumption, the best, brightest, most talented, and hardest-working people the country has to offer.)

Why, then, is the unemployment rate not 98 percent?

The answer is as blindingly obvious as it is blandly ignored in discussions of future productivity gains: It's because we want more than our ancestors had. We want not only shoes but cars for everyone, a great highway system, and easy jet travel to anywhere in the world. We like being able to take pictures of our children. We value having penicillin instead of dying of strep throat, as George Washington did. We like central heating and air-conditioning and stoves that do not require chopping wood or splitting kindling. We appreciate barramundi and mahi-mahi and Chilean sea bass. We like refrigerators to keep them in and supermarkets to buy them from. We desire not only the occasional handbill but books, newspapers, magazines with color pictures, radio, television, movies, and the internet. We think of baths, indoor toilets, underwear, deodorant, and shampoo—not to mention smartphones—as necessities of life.

This is far from a novel observation. In 1933, John Maynard Keynes blamed unemployment on "our discovery of means of economising the use of labour outrunning the pace at which we can find new uses for labour."[203] The obvious solution was finding new uses. Keynes himself got this wrong—he predicted that by the end of the century we would only work a few hours a week. It was an early version of our thought experiment, but without the "level of consumption" parameter.

The interesting thing is that, throughout the 19th century and half of the 20th, the cutoff point for employment was more or less constant at 90 percent. In 1950, the male labor-force participation rate was still over 85 percent (and 95 percent for healthy men ages 25 to 50). By 2020, it had dropped to about 70 percent. In our I-80 scenario, nobody standing in California, Nevada, Utah, or half of Wyoming would have to work. (The female labor-force participation rate is not a reasonable proxy, since before the 1950s women typically did as much work as men but were not recognized as part of the formal economy.) Today, the total official labor force is currently 50 percent of the population—everyone east of North Platte, Nebraska.

One could claim that, up until 1950 or so, the U.S. had the cutoff at approximately Reno, after which it started sliding toward the east. In 1940, manufacturing amounted to about 30 percent of U.S. jobs (people standing in Illinois, Indiana, Ohio, Pennsylvania, and New Jersey), but by 2010 it had fallen to 8 percent (those east of Punxsutawney, Pennsylvania).[204] Meanwhile, as we have seen, output per manufacturing worker increased by a factor of six. But the output of the economy overall, the GDP per capita, only increased by a factor of three. There are about 3.5 million truck drivers in the country—call it everyone standing on I-80 between New York and Parsippany, New Jersey. In many states, a truck driver is the most common single typically male job. (For comparison, there are 2.5 million waiters and waitresses.) Much has been made of the prospect of self-driving trucks and the threatened obsolescence of these jobs. But the reality of the problem is that it begins to look like any other automation process. Trucks still need to be loaded and unloaded, dispatched, and scheduled. A truck making local deliveries isn't likely to be automated until we get full-fledged humanoid robots. There are actually eight million people in the trucking industry total.

Furthermore, there are well over a million separate trucking companies. A truck owner/operator won't be put out of business by having a self-driving truck; he'll just work less for the same income, or make more for the same amount of work. The bottom line is that, even with self-driving trucks, the industry will become more efficient in much the same way that manufacturing has, with each person accounting for more freight miles.

Trucks move about 10 billion tons of freight in the U.S. each year, or 30 tons per American. If productivity doubled, we could either fire half of all truckers or move twice as much stuff with the same human effort. This has happened many times over on the technological path from oxcarts to containerized multimodal freight, and there's no reason to expect it to stop. There's also no reason to think it's something new and disruptive.

In 1800, more than 85 percent of humanity lived in what we would consider extreme poverty, Level 1 on Rosling's scale, unable to afford even shoes. Now, for the first time ever, according to the World Bank, that figure is in the single digits: 9 percent. This is because of machines, because of increasing productivity, because of automation. The sooner that the last of the starving-dirt-farmer jobs are gone, the better.

## A Cleaner, Better Breed

Around the turn of the century, I wrote an essay called "Ethics for Machines." That was when we were beginning to see the very earliest light of the alarmist wave of "superintelligent AIs will take over the world and turn us all into paper clips." I had what I considered to be a fairly cogent critique of AI as it was then practiced. By that point, I had worked in AI for about 25 years, but the field as a whole had spent 50 years attempting to solve the natural language problem: to understand, generate, and translate English (say) competently.

Now it turns out that one of the major achievements of moral philosophy in the 20th century was to come to the realization that the great piles of little rules we had accumulated in an attempt to map what was moral behavior were very much like the great piles of little rules—grammar—we had accumulated to explain how language worked. Our moral facility was very much like our language facility.

Unfortunately, AI had just spent 50 years proving that we didn't have a very good handle on language. Even after all the work we had put into them, the programs could neither understand nor produce natural, flexible, colloquial English (or any other language). So, how much time had AI researchers spent on giving the robot a sense of morality, presumably a task of the same difficulty as language?

In a word, zero. There were no books on the subject, and precious few essays. Whereas there had been a growing field of computational linguistics, with journals and conferences, since the mid-1970s, there was essentially nothing in computational ethics. When my essay grew into a book (*Beyond AI*) and got published in 2007, it was the first full-length nonfiction book on the subject.

The world of AI has changed a lot since then; indeed, it has passed over the watershed of expectations that I described earlier. Back in 2007, I felt I had to spend the first third of the book just convincing the reader that human-level AI was something reasonable to expect, and not simply science fiction. This is obviously no longer necessary.

The techie world has been agog over the past year at the abilities of DeepMind's GPT-3, a deep network like the toy one at the beginning of the chapter, but which is supercomputer-size and trained on an outrageously enormous quantity of data. It can produce fluid, coherent, lifelike essays at approximately the level of a college student or journalist on virtually any subject.

Have we solved the problem of language, then? If so, can we solve the problem of machine ethics the same way?

I'm a frayed knot. GPT-3 doesn't really understand what it's saying. It merely regurgitates a glib pastiche of material that it has seen, over and over again, in its massive dataset. Indeed, one typically has to elicit several efforts from it and pick the best in order to get the kind of startlingly good results that we have seen.

At this point, it is a good idea to understand what I mean when I say "understand." When I understand something, be it what a duck is or what the pathetic fallacy in literature is, I have a mental model of it. A mental model of something enables you to do several things:

> › Recognize it when you see it (or hear it, read it, etc.).
> › Imitate it or simulate it.
> › Predict what it will do.
> › Know how or where to cause it, produce it, or find it.
> › Know what it is useful for and how to use it.
> › Know whether it is dangerous and how to avoid it.

I can walk like a duck, quack like a duck, recognize a duck when I see one. I can draw a picture of a duck. I know much better what it is like to be a duck flaring its wings for a water landing than I did before I became a pilot. I know how to cook a duck and what it tastes like thereafter.

GPT-3 can do none of these things. What GPT-3 understands, to the extent that it understands anything, is *verbal references* to ducks. It knows when the word "duck" is likely to occur as a function of the preceding words, and what words are likely to follow. The chattering literati are apt to mistake this for understanding the duck, but I am not, because I understand the chattering literati.

The harder problem is the one of Rosie knocking over the furniture 100,000 times to learn. There simply isn't a corpus of a billion moral recognition, analysis, and decision sequences available at a level that a deep network can assimilate.

Where do we stand for trusted robots, then? Do we have to solve the deep problems of ethics that philosophers have wrestled with in vain for 3,000 years?

No, not really. What we do in the short term is make Asimov's robots, whose major rule in life is "Do what the humans tell you to do." This is not the ultimate solution to the eternal question of goodness and badness, of course. Because you are a product of evolution, your design includes the legacy of every scheming monkey, vicious shrew, and fratriphagic fish in your ancestry who beat out their brethren in an unbroken billion-year existential succession. Having the robot do what you tell it to is hardly the perfect solution; but it's good enough for a start.

One of the things that humans are particularly good at is telling others, starting with dogs and children, how good or bad they are being and how much we like or don't like what they are doing. After that, it will be a case of building up the corpus of practical and moral judgments. Almost all of this will consist of simply knowing how to do things, including making coffee. The more that robots record all these judgments and share them in a central repository, the sooner we will have a corpus on which to base commonsense ability and morality.

I spent much of *Beyond AI* arguing that if we can make robots smarter than we are, it will be a simple task to make them morally superior as well. Asimov thought so, too: "They're a cleaner, better breed than we are."[205] To build a better robot, we need only copy the "thin veneer of civilization" at the top of the human mind; a good robot is a lot simpler than a humanlike one.

If you want to understand the essence of machine ethics in the near term, simply imagine that you have managed to design and program a robot that can perform the tasks of a salesclerk in a retail store. How much extra work must you do to improve it to the point where it won't steal from the store? The answer, of course, is none.

Ultimately, this is the reason we want robots instead of people in positions of trust and responsibility, summed up in the word "reliability": The robot works 24/7; the robot doesn't take vacations or family leave or get sick. The robot will not sue you for hiring bias or sexual harassment. It will not harass your other employees, causing them to sue you. Most importantly, the robot will not steal from you. Employees steal about $50 billion from American businesses each year, causing about a third of business bankruptcies.[206]

As machine ethicist Ronald Arkin points out, "It's a low bar."

The flip side is the story of "The Sorcerer's Apprentice" or "The Monkey's Paw." In modern meta-AI studies, we have the story of the robot told to manufacture as many paper clips as possible, so it invents superweapons beyond the dreams of science fiction, and blows up the Earth to get at the iron core for raw material.

I have never been able to take this seriously. It contains an outrageous contradiction: a confabulation of a superhuman godlike intelligence and its utterly abysmal ignorance of any hint of common sense. Anyone who designed or built a machine like that would have to be a psychopath who made it on purpose, in his own image.[207]

This is simply not the way that sane people are going to build robots. The machines are not going to spring out, fully superintelligent, like Athena from the forehead of Zeus, anyway. Instead, as part of the process of building more and more capable robots, we are going to pay a lot of attention to whatever the ones we have built so far get wrong, and the act of fixing those is going to be a significant part of the overall improvement. So, at every stage, most of what the robot can do it does right.

Let us suppose that we can build, say, in 2030 or 2040, a doctor robot as good as an average human doctor. As the price of computation falls, we can combine the general practitioner with more and more specialists, until we have something with at least a superhuman breadth of knowledge. We can also improve the speed of its thinking, and very likely in the process we will learn enough to improve it toward the high end of the human range of intelligence.[208] It's not at all hard to imagine that we could have a doctor with encyclopedic expertise, instant built-in access to all the latest research, and an effective IQ of 200. Oh yes, and it doesn't get tired or make mistakes.

Even if it isn't cheaper, that is the doctor I will be going to—or, more likely, who will be coming to make a house call by downloading into my own service robot. Similarly, that's the kind of lawyer and tax accountant I'll consult. I want a robot architect to design my house, a robot investment counselor, and robot engineers to design my flying car.

Who's going to program these robots? You are. You say, "Well, I wish it could have been like this." The smarter the robot is, the more likely it will be able to understand what you wanted, whether it was possible, and how it might be possible, and then to produce a more satisfactory result the next time. Indeed, once the robots have gotten to the level of being able to understand (truly *understand*) plain English, no more programming of the kind that we do today will be necessary.

One of the reasons that we have stumbled moving from Level 4 of wealth to Level 5 is that we don't have a clear idea of the advantages it could bring. Yes, a flying car will allow you to have dinner in Paris with the same convenience you now go to Poughkeepsie, but is that life-changing? On the other hands, robots are likely to make a huge difference, whether it be a higher level of competence among professionals or an affordable housemaid for the masses, à la *The Jetsons*' Rosie. Self-driving cars, the futuristic fad du jour, are the equivalent of everyone having their own chauffeur. We could each have a full domestic staff by the end of the century. Add to that the physical productivity of nanotech, and you really do get a significant quantum jump in the quality of life.

> *We can only see a short distance ahead, but we can see plenty there that needs to be done.*
>
> —**Alan Turing**

# 15

# The
# Second
# Atomic Age

If you had given Thomas Newcomen, circa 1700, a tank full of gasoline, he would not have been able to use it for much except, perhaps, to burn his new steam engine down or blow it up. *The technology did not exist to make proper use of such a high-power energy source*. It took another century to develop tools with the strength and precision to enable the internal combustion engine. But then the knowledge and technology of the Industrial Revolution allowed that engine to grow from working models, circa 1800, to workhorse, circa 1900. Then we got the Wright brothers' gasoline-powered shop engine—and flying machines.

Similarly, the technology of the next Atomic Age will be a confluence of two strongly synergistic atomic technologies: nanotech and nuclear. The Second Atomic Age will, first of all, be the age of atomically precise machinery. And, stretching a point, we will adopt the original meaning of "atomic energy" as well. Properly developed, nanotech and nuclear fit together like chemical fuels and steel machines. Nanotech enables broadly based nuclear technology. There are three fairly straightforward ways that it will do so.

First and most obvious is isotopic separation. Isotopes are nuclei with the same number of protons but different numbers of neutrons. Because they have the same charge, the electronic structure that they impose on an atom is the same; chemically, they are the same element. That makes them extremely difficult to separate with conventional chemical techniques.

But because the number of neutrons is different, their properties in nuclear processes are profoundly different. Famously, U-235 is fissile but U-238 is not. More esoterically, xenon-134 is essentially stable, with a half-life of 58 billion trillion years and a thermal neutron cross section of just 0.26 barns, while Xe-135, differing by only one neutron, is vigorously radioactive, with a half-life of eight hours and a cross section of 2.6 million barns.

There are many examples in nuclear design where a particular isotope is much more useful than another. This means that in nuclear engineering isotopic makeup is important throughout the mechanism, not just in the fuel. Separation is expensive with current bulk technology, but it could be much cheaper with nanotech, which will build everything atom by atom. If your machines can handle an atom at a time, you simply sort them by weight, or nuclear magnetic properties. This is not to say that it would be simple or straightforward: At that scale, a radioactive atom is a bomb that will randomly go off and wreck your machinery. But nanotech gives you the basic tools to apply lots of ingenuity to the problem at the right scale.

Given current technology, isotopic separation is far too expensive to be used to clean up stuff that has been exposed to neutrons—"low-level nuclear waste"—but with nanotech that process would be straightforward. This is the capability that makes nuclear into a clean, contained technology.

The second big advantage will be nanotech's ability to build extreme or precise structures in the target nuclear mechanism itself. We would still be nowhere close to manipulating the nucleus directly; the size of the nucleus relative to the atom is about that of a golf ball sitting on the 50-yard line relative to the whole football stadium. But at the atomic scale you have a lot more control over what nuclei are present, and in what relation to each other. This means you are able to build structures that provide intense electric and magnetic fields, which can influence nuclear reactions.

It turns out that a working moderated reactor can be built using just 25 grams, less than an ounce, of californium-251.[209] But Cf-251 is a purely synthetic element, which currently is only produced in reactors in microgram quantities. It would take an extensive nuclear industry to produce it in quantity *and* an inexpensive separation process to enable very small reactors to use it. The main problem with current methods for transmutation is that the main technique we have for changing one isotope into another is bombarding it with neutrons. But, when we are trying to produce an isotope that is highly reactive to neutrons, the product nuclei are much more likely to be ruined by a given exposure than the precursor is to be properly treated. So bulk technology is simply incapable of doing a lot of the nuclear processing that we could do in theory if we could handle atoms individually.

The third advantage is simply the immense productive power of nanotech. You can imagine having your machine under continuous reconstruction, with the entire mechanism replaced with new parts every day. This cuts down considerably on the buildup of defects due to radiation, for example. (That is essentially the way that *Deinococcus radiodurans*, the remarkably radiation-resistant "Conan the Bacterium," works.) Cell-repair machines would make us considerably less vulnerable to the health effects of radiation; a happy by-product of such technology would be allowing us to live in space.

On the flip side, without nuclear energy, nanotech is drastically underpowered. Typical piston engines produce less than a horsepower per pound of engine weight. Gas turbines produce 2.8 at the size of a private airplane turboprop, and 5.8 in the GE90 airliner engine. Modern rare-earth electric motors are in the same ballpark. But a Drexler electric motor, as mentioned in Chapter 4, would produce about three billion horsepower per pound. Nuclear power densities and nanotech power needs are roughly matched, like chemical fuels and steel machines.

## Renewable Energy

*It's reasonable to expect the supply of energy to continue becoming more available and less scarce, forever.*

—**Julian Simon**

There are something like four billion tons of uranium dissolved in the Earth's oceans. That works out to be over 100 quadrillion watt-years of energy, enough to supply the current American 10-kilowatt level of power to 10 billion people for 10,000 years. Surprisingly, current technology is quite close to extracting uranium from seawater economically. The April 20, 2016, issue of the journal *Industrial and Engineering Research* was dedicated entirely to the subject, for example.

Economical extraction would be a cakewalk with full-fledged nanotech. We could put the entire world back on the Henry Adams Curve, and the only environmental impact would be to make the oceans imperceptibly less radioactive. Or maybe not even that.

Uranium is found dissolved in seawater in very tiny amounts, about three parts per billion. However, this level represents a homeostatic chemical equilibrium, rather than simply a running sum of the thousands of tons per year of uranium that the rivers have added to the oceans over Earth's history. The oceans contain four billion tons of dissolved uranium at any given time, but ocean-floor rocks contain 100 trillion tons, deposited from the water over geological time. If we took out enough to lower the concentration, it would simply start leaching back out of the rocks to maintain the equilibrium.

In other words, if we start taking uranium out of seawater and use it for the entire world's energy economy, indeed a robustly growing energy economy, the concentration in seawater will not decline for millions of years.

For 60 years, General Atomic has been making and installing research reactors, which, unlike the big Navy-derived pressurized water reactors used for commercial power generation, are safe—safe enough to be left in the hands of a drunken graduate student. Based on an idea and design of Freeman Dyson, the TRIGA reactors, as they are called, do not rely on external quenching rods or cooling. The uranium-zirconium-hydride fuel is formulated so that it has a negative coefficient of reactivity, which means that the hotter it gets, the less it supports the fission chain reaction. This is essentially the exact opposite of how an implosion bomb works. In a bomb, squeezing the atoms together makes a noncritical mass critical; in the TRIGA, thermal expansion makes a critical mass noncritical. A grand total of about 70 TRIGAs have operated over the course of 60 years without a significant accident.

The TRIGA stands as a straightforward proof that safe reactors can be designed. How small could we make one? In a neutron-reflected, solution-based, moderated re-action, for example, the critical mass of plutonium-241 is 0.246 kilograms: just over half a pound, or three-fourths of a cubic inch.[210] Cf-251, as noted above, comes in at less than an ounce. Given how dense it is, that's just one-third of a teaspoon. Reactors based on this principle, with the fuel dissolved in a liquid moderator, have a long histo-ry, going all the way back to Los Alamos and extending through the 1950s at Oak Ridge. (They are called *aqueous homogeneous* reactors; about 30 of them have been operated worldwide.) The World Nuclear Association tells us:

> A theoretical exercise published in 2006 showed that the smallest possible thermal fission reactor would be a spherical aqueous homogeneous one pow-ered by a solution of Am-242m($NO_3$)$_3$ in water. Its mass would be 4.95 kg, with 0.7 kg of Am-242m nuclear fuel, and diameter 19 cm. Power output would be a few kilowatts. Possible applications are space program and portable high-in-tensity neutron source. The small size would make it easily shielded.[211]

Nineteen centimeters is seven and a half inches. In a less extreme design, as stud-ied by Steven Wright and Ronald Lipinski at Sandia, you need about five gallons of water for the moderation and reflection, no matter how little or what kind of fission-able material you use, and the reactor core is a sphere smaller than a cubic foot.

The first full-scale reactor, then known as an atomic pile, was built at the Hanford B site, in Washington State, to produce plutonium for the Manhattan Project. When it first went into operation, it ran for three hours, began losing power, and stopped en-tirely after about a day. After a while, it started up again, by itself, and repeated the performance.

Unbeknownst to the scientists, what was happening was that xenon-135, a by-product of fission, was absorbing neutrons. Enough Xe-135 had built up to take out most of the neutrons propagating the chain reaction.

But Xe-135 is itself vigorously radioactive, with a half-life of about eight hours. So, if you leave it sitting around for a day or so, most of it is gone, and your reaction can proceed again. Luckily for the Manhattan scientists, the pile had been built twice as big as their calculations called for, so they could essentially use brute force to over-come the xenon poisoning. With a solid-fueled reactor, that's about all you can do.

A liquid-core reactor, in contrast, can have the fuel circulating continuously, and have the Xe-135 filtered out by constantly self-rebuilding molecular sorting machin-ery, along with any other by-products you may wish to remove—which means that the reactor doesn't have to be overbuilt by a factor of two. And that, in turn, means that if something goes wrong with the recirculating and filtering machinery, the reactor will quench itself.

The engineering possibilities are endless. I'm doing nothing but back-of-the-envelope rumination here, but it seems likely that, with nanotech-level engineering, the basic physics would allow for a reactor that would fit in a closet and produce on the order of a megawatt. A sanity check is provided by the fact that NASA is currently developing, under the project name Kilopower, a 10-kW reactor of about that size for use in space probes. It is built around a solid U-235-molybdenum alloy core, which is "about the size of a roll of paper towels."[212]

Nobody in their right mind would build a reactor like this for power production today (except on a space probe). It would be like buying a million-dollar computer in the 1960s just to play solitaire. But, after 50 years of Moore's Law, we do the equivalent all the time and think nothing of it. With the physical productivity of a nanotech industrial base, what would have been a ridiculously expensive way of powering a household, or a vehicle, would become perfectly reasonable. And it would come with benefits, such as putting your home anywhere, including a mountaintop in the Canadian Rockies, or on the South Pacific, or indeed surfing the jet stream at 50,000 feet. No power lines, no oil or coal deliveries, no emissions.

A home reactor providing an average of 100 kW, about right for the Henry Adams Curve in the Jetson era, would produce about an ounce of fission by-products per year. Most existing reactors running at that kind of power level don't even bother to remove the waste from the fuel assemblies.

The standard fission reactors use isotopes that can be fissioned by thermal neutrons—ones that have been slowed down to have the same energy as ordinary atoms at room temperature, or about 0.025 electron volts. The word *fissile* is used for isotopes that are fissionable by thermal neutrons and capable of sustaining a chain reaction. The cross section, or the likelihood of capturing a neutron, is much higher for thermal neutrons than for faster ones. This is why a moderated chain reaction can use a much smaller quantity of fissionable material than a bomb can; the moderation is what slows the neutrons down.

Some isotopes that are not fissile are still technically fissionable; they have to be hit with a much higher-energy neutron to fission. Uranium-238, which comprises 99 percent of natural uranium, is an example. (If U-238 encounters a thermal neutron, it tends to capture it, becoming U-239 and beginning the process that turns it into plutonium.)

If you have a source of high-energy neutrons and a chunk of U-238, you can build a reactor that doesn't use a chain reaction, and, in fact, can't run at all if the external neutrons are not supplied. This reactor would be runaway-proof and meltdown-proof, and it could use natural unenriched uranium (or the more abundant thorium), so no enrichment facilities (such as might be used to produce bomb-grade material) would be necessary. That would make it both safer and cheaper than current reactors.

Where do you get the neutrons? One option is a particle accelerator. You can't accelerate neutrons directly, but, in a process called spallation, you can throw charged particles, like protons, at materials that will give off neutrons when struck. The other way to get high-energy neutrons is by fusion, notably of deuterium and tritium. This reaction produces helium and a neutron of 14 mega electron-volts (MeV).

It turns out that you can get an interesting form of fusion, halfway between hot and cold, by pre-loading erbium with deuterium, and then bombarding it with appropriate high-energy particles. The project at NASA Glenn, where this was first experimentally achieved, refers to the process as "lattice confinement fusion," and, as you might guess, it's under investigation as a way to power spacecraft.[213]

## Assassins and Accidents

In 2006, MI6 agent (and former Soviet spy) Alexander Litvinenko was assassinated by the expedient of slipping 26.5 micrograms of polonium-210 into his tea. Po-210 is a vigorous alpha-emitter; if introduced into the body efficiently (i.e., by direct injection), one microgram is a fatal dose, causing death by radiation sickness.

How can we in good conscience recommend a world energy economy based on nuclear power if nuclear materials are so deadly in such microscopic quantities? It might help to redescribe the incident: Litvinenko was killed by Russian agents who slipped *10 million dollars' worth* of polonium-210 into his tea. Even so, it took him three weeks to die. Once MI6 figured out what had happened, the source of the attack was easily traced, because there is only one facility in the world that produces Po-210, a Soviet-era plant named Avangard, in the city of Sarov. The assassins gave it to Litvinenko using a method that would have worked with just about any poison. If they had given him the metal as a scrap of foil in an ordinary paper envelope, he could have handled it in perfect safety.

Still, polonium-210 is quite dangerous if administered internally.[214] It is rare—so rare that the Russians thought it wouldn't be identified—and expensive to manufacture. It isn't manufactured as a part of the fuel-preparation or waste-refining processes for any form of power generation. Botulinum toxin is about five times as deadly, by weight, as Po-210. And, far from being extremely rare and difficult to manufacture, botulinum toxin can be made accidentally when amateur canners don't pasteurize their jams and jellies properly. It is also widely produced and even more widely handled . . . as a beauty treatment, a.k.a. Botox.

The point is straightforward. The specific toxicity of a substance tells you nothing whatsoever about whether it is any kind of a public hazard. That depends on many factors, but among them is how much of the stuff you must handle for the purpose it is being used for. We may have an intuition that a microgram of something must necessarily be easy to misplace, but when we realize that we are talking about something very expensive and very powerful, we can begin to see that people will invest an amount of care and precaution despite—even because of—its physical size. Consider tritium, which can be dangerous because it chemically substitutes for hydrogen and, when incorporated into water, could be absorbed into bodies. But tritium costs $30,000 per gram. There are some substances that you are highly motivated to make sure just don't escape.

This is where the factor of a million in energy content can make so much difference. Suppose that you use 1,000 gallons of gas per year in your automobile. That comes to 60,000 pounds of gasoline that you must handle and pump over the 10-year life of your car. During that time, the National Fire Protection Association tells us, there will be about 50,000 fires and explosions at American gas stations. So, when you pump gas, you pay what seems to be a reasonable amount of attention to the process, and almost all the time that works fine. But there are two downsides: first, the fires, when they do happen; and second, all the time you spend paying attention so that they don't. Now, it turns out that half an ounce of thorium contains the same amount of energy as those 60,000 pounds of gas. If we but understood how to use it at that scale, you would never fuel your car at all; fuel for the car's lifetime would come built in, factory-sealed. No fires, and no worries.

The flip side of the equation is that nanotech makes it much easier to clean up any accident or cure any actual exposure. We will be building cell-repair machines to prevent and cure cancer anyway; we will be building artificial organs because they will work better than the organic ones. In the Second Atomic Age, Litvinenko would have gotten a text from his left kidney telling him that it had collected 26.5 micrograms of polonium-210, and what would he like to do with it?

## Fusion, Cold

In 2014, when I attended the 25th-anniversary conference on cold fusion at MIT, I found myself among stable, sane people with respectable scientific credentials and decades of careful experimental work. Addressing a local science-interest group on the subject, I even discovered that an acquaintance of mine had been an eyewitness to the original Fleischmann and Pons meltdown at Utah. Thirty years on, there is some evidence that the Machiavelli Effect firestorm may have begun to burn itself out. There has developed over the interim a separate subculture of LENR researchers and entrepreneurs with their own conference and publications, and a body of knowledge and techniques that has apparently improved to the point that the key experiments are now much more reproducible than they were 30 years ago.

The March 22, 2021, *IEEE Spectrum* carried the article "Whether Cold Fusion or Low-Energy Nuclear Reactions, U.S. Navy Researchers Reopen Case." It elaborates, "Scientists at the Naval Surface Warfare Center, Indian Head Division have pulled together a group of Navy, Army, and National Institute of Standards and Technology (NIST) labs to . . . conduct experiments in an effort to establish if there's really something to the cold fusion idea."[215]

The Machiavelli Effect that made cold fusion untouchable may be cooling, at least at the edges. But that's on the experimental side. The theoretical side is a different picture, recalling Sidney Harris's famous "Then a miracle occurs" cartoon.

Science advances, funeral by funeral. Sadly, after a lifetime of solid service to science, John Huizenga's obituaries mostly noted only "discredited cold fusion." But, in fact, his critique remains a fair guideline to the problems that a physical theory of cold fusion must overcome. There are three "miracles" that must be explained:

> The Coulomb barrier: The amount of energy necessary to force two nuclei close enough to fuse is much higher, by orders of magnitude, than the amount of energy that any electrochemical process can provide one atom.

> The reaction pathways: When deuterium nuclei do fuse, they almost always produce tritium and a proton, or helium-3 and a neutron, which are not detected in appropriate quantities in cold fusion experiments.

> The output: Even if, somehow, you managed to get the deuterium to fuse and stay fused in the form of helium, that pathway produces very energetic gamma rays (which are also not detected), rather than heat spread out evenly over tens of billions of atoms.

One single D-D (deuterium-deuterium) fusion event, if resulting in helium, produces 24 mega electron-volts of energy, enough to raise a typical bacterium 1° Celsius. The bacterium is big enough to contain 33 billion or so heavy water molecules. If you fused the pair of deuterium atoms in each molecule, you should get a temperature of 33,000,000,000°C, if that made any physical sense, which it doesn't. Instead, when we do force deuterons to fuse in particle accelerators, all that energy is concentrated in the one helium nucleus that we were hoping to get, and almost always blows it to smithereens.

To date, most of the theorizing about cold fusion has gone into trying to understand the first miracle, getting fusion to happen in the first place. To be fair, all of the effort in hot fusion has gone into the same phase of the problem, and we have yet to solve it.

But we would like a theory in which whenever some mechanism causes miracle one to happen, it *almost always* also causes miracles two and three to happen. The most non-miraculous explanation would be simply that they are all the same mechanism.

Peter Hagelstein champions a theory that appears to bag all the miracles with one phenomenon: quantum coherence. It is based on the idea that there is a direct quantum-mechanical coupling between some nuclear transitions and phonons in the solid state.

A phonon is a quantum of heat, in just the same way that a photon is a quantum of light. It bears the same relationship to the physical motions of the atoms in a substance that the photon does to the wave motion of the electric and magnetic fields that make up the classic description of light. By this description, a light wave is a coordinated oscillation of the electric and magnetic fields, and it exerts a force directly on the electrons of an atom in accordance with Maxwell's equations. A little wave jiggles them a little, a big one jiggles them a lot. In theory. But a century ago physicists realized that's not what actually happened. Whether an electron gets thrown out of an atom, for example, depends not on the amplitude of a light wave but on its frequency. Not its brightness but its color!

It was Einstein who won the Nobel Prize for seeing the theoretical implications of this "photoelectric effect," and he famously grappled with an intuitive understanding of it ("God does not play dice") for the rest of his life.[216] So has everyone else. Even Bohr is quoted as saying, "Those who are not shocked when they first come across quantum theory cannot possibly have understood it."[217] And half a century later Feynman summed it up with his oft-repeated quip: "I think I can safely say that nobody understands quantum mechanics."[218]

The bottom line is that the physicists of the 1920s developed a theory that, while intuitively frustrating, predicts the results of experiments very precisely, explains the structure of atoms in a way classic physics simply could not, and has allowed us to design a host of devices, such as transistors, LEDs, and lasers, which rely entirely upon quantum effects for their operation.

For example, one of the weird quantum-mechanical phenomena that makes a laser work is coherence. If you "pump" the atoms of a material to a higher energy state (itself a quantum concept), normally they will just drop back down at random times, emitting photons in random directions. But it is possible to arrange things so that the atoms are stimulated into a coherent state change, emitting all the photons at the same time, in the same direction, with the same polarization and frequency. And it happens much faster than a noncoherent process would have done.

So, if there were a coherent quantum-mechanical coupling between nuclear degrees of freedom and phonons in the atomic lattice of the palladium, it would do the following things at the same time: provide a channel for excitation in the metal to overcome a nuclear energy barrier; affect the reaction pathway by providing a wide-open "drain" for the energy along one of them; and provide a way for the energy to reappear as heat instead of hard radiation.

In addition to this intuitively appealing explanation (and in spite of a formidable mathematical exposition in his papers), Hagelstein's theory is attractive because it apparently predicted some otherwise unexpected experimental results. In one, optical stimulation of a Fleischmann and Pons–style electrochemical cell enhanced the excess heat effect at a phonon frequency associated with the passage of deuterium through the palladium lattice, but not at other frequencies.

At first blush, it seems that saying there might be a quantum coupling between phonons and some nuclear degree of freedom is indistinguishable from magic. But it's not completely insane. The phonons of interest are in the terahertz frequency range and are composed of vibrations of the atoms and the nuclei. But when you push an atom, what pushes the nucleus is the electromagnetic field of its electron cloud. Inside the nucleus, this pushes on the protons but not the neutrons, so we can assume that there is, at the very least, some dependence on intranuclear forces to move them all together.[219] It turns out that the terms in the Dirac equation that allow this coupling are relativistic, and are almost always ignored in calculations about solids that are standing still. For ordinary objects and speeds, this is perfectly fine—the relativistic corrections for a baseball pitch, for example, show up in the 13th decimal place. But maybe not for nuclei under just the right conditions.

Let us suppose, for the sake of argument, that Fleischmann and Pons were right. Or, more precisely, were right about a heat effect, though wrong about the fusion pathway. This does not mean that the energy millennium is right around the corner. To begin with, palladium is rare—it's a platinum-group metal that is used in jewelry; indeed, my wedding ring is palladium!—and that generating capacity would be fairly expensive even if the F&P cells had worked every time. The most reliably reproducible LENR experiments I know of, Mitchell Swartz's "nanor" preloaded nano-material composite wires, produce 100 milliwatts of heat from one milliwatt of electricity. He told me that the wires cost over $100,000 to make. This should be a scientific bombshell; it is a commercial dud.

There is one cold fusion effort that, for whatever reason, largely avoided the Machiavelli Effect. Operating out of the Navy labs at SPAWAR, the team behind it had a somewhat novel approach to the electrolysis, not only loading deuterium into their electrodes but co-depositing more palladium at the same time. Then, instead of doing calorimetry to look for excess heat, they looked for radiation. Using standard methods, they found it in unambiguous form. As a result, they have 50 peer-reviewed publications in 12 different scientific journals, and two patents, which are extremely hard to come by for cold fusion researchers.[220]

Their cells specifically produced high-energy neutrons, such as might be used to run a chainless reactor burning U-238. This is, in fact, the research that led to the NASA "lattice confinement fusion" experiments mentioned above, and the Navy-led multi-lab research effort just starting.

Although at present there are several companies with apparently promising developmental results attempting to commercialize various forms of LENR reactors, there has been a steady stream of such attempts since 1989 and nothing useful has resulted. There is a huge difference between demonstrating an effect in the lab and making it reliable, powerful, and, above all, inexpensive enough to be economically viable.

And that will likely not happen until we gain a solid theoretical understanding of what's going on in atomic-level detail, *and have the technology to manipulate the conditions at will.* In short, we need nanotech. If there really is a back door in obscure quantum mechanics between nuclear degrees of freedom and the solid state, it seems quite likely that the configurations of atoms that unlock it are produced rarely and at random by bulk manufacturing methods. With nanotech, we could first analyze what was going on at the atomic scale in detail, and then reproduce it reliably.

The mind-boggling, Failure of the Imagination implications of a new physics backdoor are not so much that we might be able to reproduce a couple of decades worth of laboratory curiosities; they are, instead, that a solid understanding of what is going on could give us a powerful handle to control, at the atomic scale, all the different forms of nuclear energy we already know about.

For example, we don't know of a way to fuse two nitrogen nuclei to produce silicon. But if we could do so, it would produce about one million times as much energy per weight as gasoline. Fusing the approximately one gram of nitrogen contained in one liter of air—a good hard sigh—would produce a quarter teaspoon of sand as ash, along with enough energy to run George Jetson's 1,341-horsepower flying car for 13 hours.

If we could master just this one reaction, no machine on Earth would ever need fuel.

The real tragedy of the cold fusion episode is not about fusion, or quantum mechanics, or indeed energy per se. It is that, over the course of the Great Stagnation, the funding-bureaucracy/academia complex has lost the ability to do purely experimental, observation-driven research. Failure of Imagination is now the law of the land.

## Fusion, Hot

In 1954, Lewis Strauss, head of the Atomic Energy Commission, famously predicted energy "too cheap to meter." He was speaking of power not from fission but from fusion. With people, including the commissioners of the AEC, telling them that fusion was right around the corner, the science fiction writers of the 1950s can be forgiven for over-expecting it. After all, we did produce fusion in hydrogen bombs, and fusion happens in the Sun, converting simple, plain, ordinary hydrogen into completely innocuous helium. This sounds like a dream power source. Yet, though we've spent roughly a billion dollars a year in the search for fusion, it remains stubbornly the power source of the future. (The schedule for first plasma in Europe's enormous ITER experimental fusion reactor has been pushed back to 2025, and adding tritium to the fuel mixture, necessary for power production, to 2035.) If there was one really major failure in forecasting among the technologically literate science fiction writers of the '50s, fusion would be it.

To understand why, we must start with the Sun. Fusion only occurs in the solar core, about 1 percent of the volume of the total star, where conditions are really, really extreme. The temperature is $15,000,000°$ Celsius; the pressure is 265 billion atmospheres; and the hydrogen plasma is compressed so much that one gallon of it weighs 1,200 pounds. (Under Earthly conditions, 1,200 pounds of hydrogen would fill a balloon big enough to lift seven tons.)

Under these ridiculously extreme conditions, hydrogen will fuse into helium in a three-step process. The first step, in which two protons (hydrogen nuclei) fuse and emit a positron to form a deuterium nucleus, takes an average of a billion years to happen to any given proton. The second step takes four seconds, and the third takes 400 years. But the overall sequence is a very slow reaction, and that's a good thing. If the Sun burned its hydrogen as fast as a hydrogen bomb, it would all have been gone billions of years ago! The Sun is a barely burning bed of coals, well banked against the long interstellar night. But this, in turn, means that a volume of solar core equal to the displacement of your Subaru's engine (and weighing 660 pounds) only produces half a watt of power.

The reason we are able to achieve fusion at all, in an H-bomb, is first and foremost that it isn't plain hydrogen being burned—it's deuterium, tritium, and lithium, already one or more steps along the reaction sequence in the Sun, and in particular well past the billion-year step. The other major difference is that the fission bomb used to ignite the thermonuclear reaction isn't at 15,000,000° Celsius, it's at 400,000,000°, so hot that it's not white-hot, or even UV-hot, but X-ray-hot.

Shrouded in the mists of history and secrecy is the fact that even then it's quite difficult to make the fusion go. As part of the Manhattan Project, the Los Alamos scientists produced two completely different kinds of fission bombs in just a couple of years, starting from scratch; both designs worked on the first attempt. The same (more or less) scientists worked for another five years after the war, basically to prove that all their best attempts at a fusion design wouldn't work. It wasn't until 1951 that Stanislaw Ulam had the key insight that led to the ingenious design that ultimately did.

In simplest terms, here's how: There are actually four explosions. First is the chemical explosive that compresses the plutonium core, which causes the fission explosion (number two) of that core. This is designed to produce most of its energy in the form of X-rays, which are directed to heat up a sleeve of plastic (e.g., polyethylene and/or polystyrene). The amount of energy deposited this way causes the sleeve to explode (number three), with a violence far greater than could be achieved by a chemical explosive, and it is that explosion which compresses and heats the fusion fuel to the point of explosion (number four) inside the sleeve.

Even the first-stage chemical explosion, tiny in power compared to the nuclear ones, would destroy the device. One of the bomb's major feats of design is the way that the falling-domino sequence of successive explosions happens in the short microseconds while the bomb is in the process of converting itself into incandescent plasma. There is no hint of a ghost of a chance that a reactor using anything even vaguely like this process could survive even a millisecond of its own operation.

As of today, the only way to produce net energy from fusion with existing technology is to use actual thermonuclear bombs in a chamber thousands of feet across, probably underground. The chamber would be filled with steam, which the bombs would heat up, and this heat would be drawn off to run turbines of conventional design. Two explosions a day would produce as much continuous power as a standard nuclear plant. This was seriously studied in the 1970s, under the aegis of Project PACER; although challenging, there were no technological showstoppers. However, the requirement of producing a steady supply of hydrogen bombs was a major liability, and the economics were not as good as conventional fission reactors.

Fusion was over-touted by people like Strauss, over-expected by science fiction writers, and always seemed to be just a few decades in the future. But the underlying technology has improved considerably since the 1950s. The gargantuan government-sponsored ITER experiment in Europe is already obsolete; modern superconductors make the same capabilities possible in a structure the size of a single-family home. There are now several fairly well-funded startups researching various design approaches. The Fusion Industry Association lists 25 such companies on their web-page, along with others who consider themselves part of the supply chain. I would not be surprised to see one or more of them produce something interesting in the coming decade.

Even if we do manage to produce a magnetic or inertial confinement machine, fusion remains a process that produces copious radiation and neutrons. Of the typical hydrogen fuels, the easiest to fuse is a mixture of deuterium and tritium (D-T), needing only 100,000,000° Celsius. But tritium is extremely expensive, and the process produces 80 percent of its energy as hard-to-handle high-energy neutrons.

Somewhat better is pure deuterium (D-D), which is nonradioactive and is a constituent of ordinary water, and thus economical. It needs an ignition temperature more like 400,000,000°, but only 35 percent of the energy comes out as neutrons.

It remains a holy grail to execute a form of fusion that produces most of its energy in the form of alpha particles (helium nuclei), which can be converted directly into voltage by various electromagnetic or electrostatic mechanisms. There are several projects attempting to do this, among which the most common is the proton-boron-11 fusion/fission one, in which the proton (i.e., hydrogen nucleus) fuses with the boron to produce a highly excited state of carbon-12, which promptly breaks up into three alphas. This is about 10 times as hard, temperature-wise, to do as D-T fusion, up in the billion-degree ballpark. There is a similar proton-lithium-7 reaction that is being looked at as well.

TAE (formerly Tri-Alpha) in California, funded by Paul Allen, among others, and LPP Fusion in Lawrenceville, New Jersey, for example, have been attempting the reaction with plasmas confined in magnetic fields. Another way to get kilotesla fields is with a petawatt/picosecond laser pulse. A recent simulation study at New South Wales has a laser trap and ignition pulse touching off a 14-milligram sample of a stoichiometric mixture of hydrogen and boron-11, and capturing 1 gigajoule (GJ) of pure electricity—no boilers, turbines, or generators necessary—at an energy density of 71.5 terajoules per kilogram.[221] These approaches are room-size machines costing millions, as opposed to machines of the ITER variety, which are skyscraper-size and cost billions. Some of these new projects could produce a car-size power unit, given a nanotech technology base. Whichever particular pathways we take, or physics we discover, the possibilities of the Second Atomic Age make all we have done so far seem minuscule by comparison. We will have been like a boy playing on the seashore, diverting ourselves in now and then finding a smoother pebble or a prettier shell than ordinary, whilst the great ocean of energy lay all undiscovered before us.

> *Impossibilities no longer stood in the way. One's life had fattened on impossibilities. Before the boy was six years old, he had seen four impossibilities made actual—the ocean-steamer, the railway, the electric telegraph, and the Daguerreotype; nor could he ever learn which of the four had most hurried others to come. He had seen the coal-output of the United States grow from nothing to three*

*hundred million tons or more. What was far more serious, he had seen the number of minds, engaged in pursuing force—the truest measure of its attraction—increase from a few scores or hundreds, in 1838, to many thousands in 1905, trained to sharpness never before reached, and armed with instruments amounting to new senses of indefinite power and accuracy, while they chased force into hiding-places where Nature herself had never known it to be, making analyses that contradicted being, and syntheses that endangered the elements.*

—**Henry Adams, *The Education of Henry Adams***

# 16

# Tom Swift and His Flying Car

*Every city in the United States will have a landing-field and hangars for airplanes, as well as mechanics to care for them. Whether this is to be a private or public enterprise lies in the hands of the people handling such things. Much could be said for either type of establishment. The thing must come; it is as logical as one, two, three. There are some, perhaps, who remember the roars of derision which went up when the first automobile garage was established in their town. Such a thing was visionary—there would never be enough machines to make it pay!*

—**Sweetser and Lamont,** *Opportunities in Aviation*

As we have seen, a real, parked-in-your-garage, three-vehicle-solving flying car, VTOL and all, is feasible with the technology that we have today. If we had continued to advance the relevant technologies at pre-1970 rates, it would be affordable to a fair number of people. It would probably have visible wings, rotors, or ducted fans, use a turbine/generator for power, and cruise at less than 250 knots. It might or might not fold up to some extent to be driven on the roads. It would be compact, if not cramped, for the same reason that today's aircraft and even cars are, in a bid to reduce air resistance for fuel efficiency.

But it is high time for us to slip the surly bonds of the Great Stagnation and imagine what a flying car might be with the technology of the Second Atomic Age.

Our greatest lack, of course, is power. Nothing else so brightly illuminates the gulf between the ergophobic religion that regards energy as some shameful original sin and the can-do, "whatever it takes to get the job done" attitude of the 1950s. Energy poverty is one of the major things separating us from the future we were promised. This was not done to us from without; we impoverished ourselves from within. But that means that it still lies within us to change our minds, regain our birthright, and dance the skies on laughter-silvered wings.

The airliner-speeds trend chart (see Figure 6) tells us that, if not for the Stagnation and subsonic economic rut, airliners would be doing about 16,000 miles per hour by now. This is, for all intents and purposes, space travel; inside the aircraft you would only feel about a quarter gravity. Outside the aircraft you would like the air to be as rare as possible, since the drag would be 25,000 times what you would feel doing 100 miles per hour at sea level.

Perhaps you remember the "daytrip effect" hump in the value-of-travel graphs (see Figure 34), which started at about 50 miles for a car and stretched out to a few hundred for a fast jetcar. At 16,000 miles per hour, the daytrip hump covers virtually the entire world.

Go ahead and laugh, but just remember that this is what Alan Shepard did almost exactly 60 years ago, as of this writing. The one difference you absolutely must have for your convenient hypersonic runabout is that there is no way it works for you using any chemical source of energy.

## Electric Jets

In the model-aircraft world, "electric jet" means a ducted fan powered by an electric motor. Modern rare-earth magnets and other innovations have made possible electric motors at the same power-to-weight ratios as turbines, even at model sizes. Full-size superconducting ones could have better ratios yet. Electric motors would allow considerable optimization of the placement and control of thrusters, perhaps enough to have separate lift and cruise motors.

The Schübeler DS-215-DIA HST electric ducted fan can produce 56 pounds of thrust from 15.6 kilowatts of electric power (21 horsepower) in an eight-inch package.[222] Seventy-five of them could lift your flying car nicely (if noisily). They would need 1,170 watts of power. Alternatively, if you used wings and a runway, you'd only need about 10 of them and 150 watts. The reason to do this would be to give your flying car many small fans instead of a few big ones. This makes the vehicle more reliable, because you can afford to lose a few; more efficient in flight, because you have much better control of the pattern of airflow around it; and quieter. NASA's X-57 Maxwell electric airplane works this way, for example.

The major problem with electric motors is providing the electricity. Batteries are horrible from a weight perspective. When my plane is filled up, the avgas, the fuel I use, weighs 360 pounds. The same usable energy in lithium-ion batteries would weigh about three tons (and cost $170,000, without charging circuitry).[223] At that weight, I literally could not get off the ground. Furthermore, lithium ion batteries are notoriously prone to spontaneous combustion. At least two of the new electric aircraft companies have had experimental vehicles destroyed by battery-caused ground fires.[224]

There is an inescapable problem with chemical batteries, which, ironically, gets worse as the batteries get better. If you crash in a gasoline-powered plane, the tanks may rupture, and the fuel may leak, mix with air, and catch fire. That's bad enough. But rupture a high-energy lithium battery and the energy is all right there, no spilling and mixing necessary. In one Tesla crash, firefighters sprayed 30,000 gallons of water on the car for four hours before giving up and letting it burn itself out.[225]

One alternative to batteries is to use fuel cells; like batteries, these have been steadily improving since the 1960s. At the bleeding edge today, custom-built ultralight cell stacks (at $10,000 per kilowatt) can produce a kilowatt per kilogram; this is essentially piston-engine power density. The engine in my plane weighs about 250 pounds; the equivalent in fuel cells would be about 300. This is good enough for conventional airplanes (or even helicopters), but not compact VTOLs. Their main problem is that they use pure hydrogen, which is remarkably difficult to store and handle. Hydrogen infiltrates metals, causing embrittlement and making seals problematic. It is also extremely light, requiring huge containers for small amounts. One possible solution is to use ammonia—NH3. Use the hydrogen and release the nitrogen back into the air, of which it is already the major component. Liquid ammonia contains more hydrogen per gallon than liquid hydrogen does; it is already one of the most-produced industrial chemicals, with an extensive pipeline system and a price per energy in the same ballpark as fossil fuels. During the 2010s, fuel cells that can use ammonia advanced from the laboratory to commercial deployment. As I write, they are still too heavy for vehicles, and are used for remote generation and power backup. However, in the coming decade that could easily change.

As of 2021, however, the best technological choice to power an electric-fan VTOL is an optimized turbine/generator. Surprisingly enough, these can be bought off the shelf—they are used as auxiliary power units (APUs) in airliners. You can have a turbine generator plus electric-fan motors for less than the weight of a direct-coupled piston engine. But within the next few decades fuel cells will probably be light enough, and a lot quieter, too.

Once we get to our Second Atomic Age hypersonic runabout, you will see that we have gone from the sublime to the patently ridiculous. Alan Shepard's Mercury Redstone carried 50,000 pounds of fuel and oxidizer. Rockets are very inefficient—but remember that 50,000 pounds of fuel is also what it takes to drag a 747 through the atmosphere halfway around the world. That was about seven terajoules of energy. And yet the actual energy in the orbital speed of your 3,000-pound car would be about 50 gigajoules, 140 times less. If you could extract and use it efficiently, your trip should cost you a tenth of a gram of uranium, costing about 15 cents.

There is another meaning for "electric jets": Suppose that there was a different way to push air than by swinging blades through it. It might be possible to apply a small force to a large quantity of moving air without the inefficiency incurred with a fan. It turns out that such a thruster exists, if it can be engineered into useful form. It works by spraying electric charge into the air, creating ions, and then subjecting the charged air to an electric field, which exerts a force on it. Such "ionic fans" have been laboratory curiosities and science fair projects for decades.

In 2013, Steven Barrett and Kento Masuyama at MIT did what may have been the first rigorous evaluation of ionic thrust as a propulsion method.[226] Their results were surprising: They found that an ionic thruster could produce over 100 newtons (22 pounds) of thrust per kilowatt. For comparison, the electric jets mentioned above use about 6.25 kilowatts to produce 100 newtons. Your 3,000-pound car could be rendered VTOL with a 180-horsepower motor. Unfortunately, this doesn't solve the VTOL dilemma for us; the high efficiency depends on moving a larger volume of air more slowly, just as with helicopter blades.

Ionic thrust is far from ready for prime time. It involves spraying tens of thousands of volts of electric charge into the passing air from bare conductors. Failure modes could be spectacular; imagine the *Spitzensparken* when it rains! Note that five years after the original research, Barrett and team flew the first ever powered aircraft with no moving parts—indoors.

Furthermore, high-power high-voltage power supplies tend to be quite heavy in current practice. But that would almost certainly go away with Second Atomic Age technology. Once we get an appropriate power source, there is one place where ionic thrusters are just the thing. In the stratosphere, where we want to be with our hypersonic runabout anyway, the air is thin, cold, and dry. What's more, we don't need to worry about ionizing the air, because at those altitudes the Sun already does that. An added bonus: We would be helping the ozone layer, rather than hurting it.

One of the ways that we are hoping to be able to harness the energy of the nucleus is to stimulate beta emission. But beta emission is nothing but spraying high-energy electrons—exactly what the ionic fan needs in the first place. So, if we are very good and leave cookies for Santa, ionic fans may just work as the thrusters we need for high-altitude hypersonic flight.

Coming back down to Earth and going slowly for takeoff and landing, a related aeronautical feature that would be of considerable use is circulation control. This involves sending sheets of high-pressure air out of slim slots at selected spots on the wing. You can buy such a thing yourself today, in the form of the Dyson bladeless fan. The Dyson uses its high-pressure sheet to entrain 16 times as much air as the sheet itself through a hoop with nothing in the middle.

Circulation control can significantly improve lift-to-drag ratios, give the effect of high-lift devices with no (external) moving parts, and improve the effective aspect ratio of a wing. The absence of moving parts would be particularly valuable for telescoping or folding wings. A wing with ducts and slots that were merely part of its shape could be made much lighter than one with hinges, flaps, linear actuators, and all the other panoply of current-day high-lift devices.

Circulation control (and its earlier version, known as "blown flaps") has been tested by NASA and various aeronautical engineering departments since the 1970s, and is a continuing area of active research. Some recent work by Mark Moore of NASA Langley has centered on the novel use of circulation control in a propeller duct. The issue is that a different configuration for the duct is desirable for low-speed, high-thrust regimes, such as liftoff, and for high-speed, low-thrust regimes, such as cruising. Circulation control might let the same duct serve both purposes efficiently.

## Your Flying Car

*If a man can write a better book, preach a better sermon, or make a better mousetrap, than his neighbor, though he build his house in the woods, the world will make a beaten path to his door.*

—**Ralph Waldo Emerson**[227]

Do we really need wings? In the 1960s, NASA tested a series of wingless aircraft called "lifting bodies" as possible designs for reentry vehicles for spacecraft. (See Figure 44.) The first of the series, the M2-F1, was approximately car-sized, built of

Figure 44:  NASA lifting bodies.

plywood over a steel frame, and could glide at 100 knots, carrying a pilot. (In many of the glide tests, it was towed by a hot-rodded Pontiac.)

One difference between small planes and small cars that people rarely think about is that when you park your plane at the airport, you are well advised to tie it down. Sitting light without passengers and cargo, high off the ground for propeller clearance, it can be tossed around and wrecked by a gust of wind. No one even thinks of tying his car down. Even a car that only weighs a ton is nowhere near as likely as a small airplane (which also weighs a ton) to catch gusts with disastrous results.

The average car today has a ground footprint of about 100 square feet. Let's assume that you can use modern high-tech materials (including a turbine for an engine) to make your flying car one ton of machinery with 1,000 pounds of payload (typical for a small plane). Using multiple small thrusters, circulation control, and playing with the flying car's shape a bit, so as to call the whole business a lifting body, the craft has a wing loading of 30 pounds per square foot. That is high for a small plane (Piper Cub comes in at six, a Beechcraft Bonanza at 25) but low for a jet (Cessna Citation comes in at 40, a Boeing 747 at 150). You would be taking off and landing at about 100 knots.

By itself, such a lifting body represents a wing of such a low aspect ratio (width to length of the car, in this case) that it's highly inefficient. At the very least, you need air curtains and boundary layer control to steer and contain airflows around the sides. In fact, in *Childhood's End*, Arthur C. Clarke envisioned exactly this as the technological basis of "the ubiquitous little air-cars [that] washed away the last barriers between the different tribes of mankind." As he described them:

> The ordinary private flyer or air-car had no wings at all, or indeed any visible control surfaces. Even the clumsy rotor blades of the old helicopters had been banished. Yet Man had not discovered anti-gravity. . . . His air-cars were propelled by forces which the Wright brothers would have understood. Jet reaction, used both directly and in the more subtle form of boundary layer control, drove his flyers forward and held them in the air.[228]

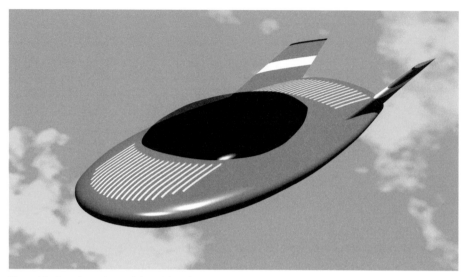

**Figure 45:** A mash-up of several circa 1960 flying-car designs, ranging from Gerry Anderson's Supercar to a Curtiss-Wright marketing department concept, with a nod to the lifting bodies and, of course, *The Jetsons*.

There are a few differences in detail between what Clarke sketches and what we might actually be able to build along these lines. It won't be a "little air-car," at least compared with a ground car (although it would be smaller than an airplane of conventional design). For one thing, our flying car needs to be big to interact with enough air to fly. For another, it needs to be light, so for use on the ground you will want crumple zones instead of heavy internal guardrails, in case of collisions.

But we might want a smaller car specifically to raise the wing loading. A fully loaded 747 has a wing loading of over 150 pounds per square foot, which makes it require higher takeoff speeds and longer runways, but also makes it much less susceptible to turbulence. Your 3,000-pound car would only need 20 square feet of wing; that's less than half the footprint of a Smart car. Alternatively, you could simply add a few tons to the weight of a full-size car.

All that said, it seems likely we could take a car shaped like this and make it take off and fly. You can throw out the power-versus-speed chart, though; all that forced air for the blown flaps doesn't come for free. Budget for a 1,000-horsepower turbine, or fuel cell, or atomic battery, or whatever it is we are going to call it.

Have we let ourselves in for the VTOL dilemma again? Maybe not! The two main sources of drag vary oppositely with speed. Parasitic drag, things like skin friction and turbulence caused by your passage, goes up as the square of your speed. But induced drag, which is caused by producing lift, goes down. Bigger wings mean more parasitic drag and less induced drag; smaller wings, vice versa. Most airplanes are designed so that the sum of the drag terms is lowest at the intended cruising speed.

Not so our flying car. It has a high wing loading, high induced drag, but low parasitic drag. So it's optimized for higher speeds. All that surface blowing and those air curtains become less and less necessary the faster you go. The power can be diverted back to thrust. At 400 knots, all you need from your car shape is a lift coefficient of 0.06—and that's easy, even from a wing with a very low aspect ratio. In real life, low-aspect-ratio wings are found on very fast aircraft—think of the Concorde. The NASA lifting bodies, air-dropped and rocket-powered, got up to 1,000 miles per hour. So, your 1,000 horsepower might just pay off in high speed after all.

*It is a possibility.*

Can we go the whole hog? Put in down-blowing fans in the trunk and hood areas for true VTOL? In theory, yes. There is enough thruster area for 1,000 horsepower to lift the car. The Airgeeps of the 1960s were designed along these lines. Is there room for anything else, including the car's engine?

Surprisingly, the answer is probably yes. With the Schübeler electric jets discussed above, you only need about 20 square feet of footprint (out of the 100 in a typical car) to lift 3,000 pounds, i.e., 54 of the jets, drawing 850 kilowatts. In practice, you might use a few more—you could derate the jets to 10 kilowatts and use 64 of them at 48 pounds thrust each and a power draw of 640 kilowatts. In either case, there is plenty of footprint left over for engine, passenger compartment, wheels, and so forth. The Doak VZ-4 put 1,000 horsepower into 50 square feet of lift-fan area; a similar area in our car would leave us 50 square feet of floor space for passengers, engine, and so forth. And, if that doesn't allow you to fit all you would like, you can add another 50 square feet and still have the footprint of a Ford F-150 pickup—or a 1959 Cadillac.

You could build a car like this, VTOL and all, today, and power it with batteries. And it would fly. But it would only fly for about 10 minutes. With current technology, pretty much the only power source that would work would be a turbine.

So you can beat the VTOL dilemma as far as speed is concerned, but you can't avoid it altogether. The weight and volume of your electric jets (say 300 pounds) and louver-closing arrangements for faring at high speeds could have been useful as both load and cargo space. If nothing else, you could have used that capacity for fuel; VTOL and range are hard trade-offs. The Doak VZ-4 had only one hour of flight time. Going VTOL probably makes a two-seater out of what would have been a four-seater, had you used a mile-long runway. Take your choice.

Electric fans with the power-to-weight ratios and reliability that we need postdate the 1960s, and there are reasons to prefer one to a direct-coupled turbine (either shaft or air) for the fans, even though the latter is probably a bit more efficient. First is the much faster control reactivity of the electric fans, which translates into much better control in hover. Second: You can afford enough batteries to power you down to a hard landing, instead of a horrible crash, in the case of a turbine failure on takeoff. Third: When we get around to building the infrastructure, we can transmit power via microwave to flying aircars, and cut down the amount of fuel they have to carry.

Today, your flying car could be turbine-powered, but it would still run on fossil fuels. That means you would have a trade-off between VTOL and range or load. With nanotech, even if we are still thinking in terms of oxidized chemical fuels, this compromise largely disappears. Full atomic control means that we do not have to thermalize chemical potential energy to use it. This, in turn, means that we don't have to strain the bounds of material science to accommodate ultrahot temperatures in your engine. It also means that we get three or four times the usable energy from a given quantity of fuel, instead of sacrificing it on the altar of thermodynamics.

Hydrogen has three times the energy-to-weight ratio of gasoline; use it in a fuel cell to produce electricity directly, or in a nanotech power mill with the same function, and you get a further factor of three or so. The Cirrus family-size jet carries 2,000 pounds, a full ton, of jet fuel; this would be cut down to, say, 200 pounds in hydrogen for a similar 1,000-mile range. The bleeding edge of current practice is fuel cells with the power-to-weight ratio of piston engines, and hydrogen storage that is heavier than the same energy in gasoline. You could do a usable convertible with that, but the other drawback is cost, which is still quite high.

Nanotech could be expected to ameliorate all three of these factors, producing an affordable VTOL car. It's probably worth mentioning that, with nanotech, producing the hydrogen from virtually any energy source at all is trivial, and compressing, cooling, and storing it would be considerably easier than with current-day techniques. Even so, storing energy as hydrogen, while considerably lighter than carbon-based fuels, will likely be just as bulky. Of course, for flying machines it's mostly the weight that matters.

The next big advantage for nanotech aircraft is the weight of engines, including both the fuel-burning power mills and the electric motors that will push the air. As we pointed out in Chapter 4, these are so small as to be negligible. Structural weights can also be smaller. Putting the two together, you get a significant ability to make powered, morphing structures where today you would have a rigid one. This would include both wings and propellers; extensible propellers or even extensible ducted fans would be possible.

You will not see propellers or ducted fans, however. Instead, designs will likely be dominated by many smaller impeller devices in arrays that are robust to any individual failure and that are responsive to control inputs, interacting with the air in regimes small enough that sound and turbulence are avoided.

This means we could make a compact vehicle that is VTOL but takes off with the whoosh of the wind through the trees instead of an earsplitting roar. The trick, for the designer, is to keep the entire downstream jet subsonic and as close to laminar as possible.

## Autopilot

Even my 40-year-old airplane has an autopilot that navigates me to a destination. The mechanical arrangements that allow a plane to fly a fixed heading and altitude have been around since the 1930s. The part that involves flying to a destination is a retrofit to a GPS. Modern commercial autopilot systems (not mine) are even capable of landing the plane.

In fact, you can get the Garmin Autoland system as part of high-end 2020-model private aircraft, such as the Piper M600 turboprop or the Cirrus Vision jet. In an emergency, all you have to do is press a big red button, and it will alert air traffic control, plot a course to the nearest airport, fly there, and land. While the Autoland system is for emergency use, it gives you a feel for what the state of the art is capable of.

The single drawback to autopilots is that they sometimes fail, and you have to be ready to take over in a hurry. As a pilot, the very first thing you need to learn is how to turn your autopilot off. Mine has an emergency kill switch literally under my thumb on the control yoke, a power switch on the control panel, and a circuit breaker, in addition to the controls on the unit itself.

As with self-driving cars, the question isn't when and whether the autopilot can be made perfectly reliable. It can't. The question is when and whether it can be made more reliable than a human pilot. Can we have autopilots in our flying cars that make them at least as safe as the cars we drive on the ground today?

Over the past decade, the answer to this question has changed from "very likely" to "obviously." Which is a very good thing, because the average person isn't a good pilot. As I've discussed previously, piloting is harder than driving a car, because we don't have the instincts for flight the way we do for navigating on the ground.

Furthermore, the closer you get to a VTOL flight regime, the harder flying is. Your author, in addition to getting a pilot's license while researching this book, took the obvious step of acquiring a remote-control quadrotor (i.e., VTOL) drone and trying to fly it. The drone is significantly harder to fly without crashing than my airplane, although limited space, displaced point of view, lack of kinesthetic feedback, and unfamiliar/el cheapo controls surely aggravated the difficulties.

The history of experimental aviation is littered with the wrecks of VTOLs crashed by experienced test pilots. But it is possible to build an automatic pilot with a complete understanding of the dynamics of the flight regime and the capabilities of the craft. And this could be built with direct connection to an artificial sensorium, which tells it all about the patterns of airflow around the machine, as well as things that a human pilot sees on his instruments, like attitude and airspeed. This could be far safer than any human pilot, much less an amateur one. Perhaps surprisingly, that is the piece of technology we already have. Commercial airplanes have had steadily improving autopilots for decades, to the point that, on the average flight, the human pilot isn't usually strictly necessary. And the autopilots now built into small hobby drones are more sophisticated than anything in the air at the turn of the 21st century.

## Highways in the Sky

There are on the order of four million miles of roadway in the U.S., of which about 2.6 million are paved. The Interstate system comprises about 50,000 miles of this. How much more usable right-of-way would we have if we used the sky instead?

The contiguous 48 states have an area of about three million square miles. There are about 10 miles of altitude accessible to aircraft. (Two of these are safe for non-pressurized airplanes with normally aspirated engines—you and your engine get breathless at about the same point.) Under current regulations, airplanes need 1,000 feet of vertical separation and five miles of horizontal separation. These regulations presume a technology where the pilot is looking at an altimeter and controlling his altitude manually (indeed, flight levels are defined by barometric pressure, not actual height), and he is maintaining horizontal separation by verbal interaction with air traffic controllers. Physically, planes can safely fly much closer together if they are going the same speed and direction; think of geese flying in a V-formation. But let's take the current regs as a point of departure.

If you merely parceled out the airspace in lanes at these separations, you would have on the order of 30 million miles of high-speed highway—600 times the length of the Interstate system, and more than 11 times that of all paved roads.

This would mean that there is room for six million aircraft in the air at one time (one million of these craft being non-pressurized). If you cut the horizontal separations to a mile—which wouldn't be dangerous given good electronic controls—you'd have 150 million miles of lanes and the same number of aircraft in the air at once. That's more than there are cars on American roads.

Moulton "Aerocar" Taylor was, of course, right: The existing air traffic control system would completely break down long, long before that many private planes got into the air. The ATC is surprisingly antediluvian for the 21st century. In an era when the distributed control of the internet routes and switches literally trillions of packets of information *per second* all over the planet, each interaction with ATC is done by voice in English with a live human being. This makes for a system that is often running at the outer edge of its capabilities. Recently, a disgruntled employee cut some cables in the Chicago Center and set fire to the building.[229] This created a 91,000-square-mile hole in the ATC system. More than 1,000 flights were grounded while controllers scrambled to reach alternate facilities as far away as Kansas City.

Besides being somewhat fragile and running near capacity, our current ATC system is not easy to expand.

Self-evidently, it is also incapable of handling the dramatically higher traffic burden that widely owned flying cars would involve. But centralized control isn't necessary for complex traffic patterns; you don't have a "ground traffic control" when you drive, much less when you walk, even on a busy sidewalk. (The scenario in Chapter 9 underscores how absurd a "GTC" would feel to an auto driver.) Nor is it necessary for flight. Watch a flock of birds land on and take off from a field, or watch a seagull-flock tornado doing a continuous rolling dive on a patch of delicious flotsam. It is clear that no human ATC could handle such collective flight patterns, but it is also clear that they are possible with purely local distributed control, using senses and reflexes not impossibly more capable than our own.

When you fly on a clear night, you see quite a few more airplanes than you do in the daytime. That's because each one has big, bright anti-collision strobes, which are visible 25 miles away. With car-density traffic, such strobes would make the sky a confusing Las Vegas-like light storm, but they could be replaced with radar-frequency beacons that encoded your car's position, altitude, and course. Your autopilot could read these all in real time and know its situation vis-à-vis every other aircraft nearby— and it would know this at least as well as the seagulls do, but on a scale 1,000 times as large. We need, roughly, a factor of 20 for our car's speed (500 miles per hour as opposed to 25), leaving us with a factor of 50 to account for being slower to turn than the seagull. Since we are not intending to be dogfighting, but just merging into and out of traffic flows, this should be plenty.

In the roughly three million square miles comprising the contiguous U.S., there are, give or take, 200,000 cellphone towers. That comes out to one for every 15 square miles, for an average distance apart of four miles. At cruising altitude, you are in view of 10–100 of them most of the time. Equip each one with a box that looks for and talks to all the aircraft in its area, very much as the cell tower already does for pocket phones. Add that to GPS and your flying car can always know exactly where it is, and where every other aircraft in the same state is, to within a foot. Add that to inter-car communication and all that remains is to work out the rules of the road.

It turns out that something like this has been implemented for the North American airspace under the somewhat opaque acronym ADS-B, for Automatic Dependent Surveillance-Broadcast. Instead of using the 200,000 cellphone towers, it uses 634 government-run ground stations. These provide coverage over much, but not all, of the continental U.S. By 2020, all aircraft operating in certain high-traffic areas (essentially the same ones that require a radar transponder today) were required to be equipped with ADS-B Out, consisting of a GPS receiver and a transmitter that broadcasts the plane's position and altitude to the system. Whenever that data is relayed— about once a second—the system transmits position information back to the airplane about traffic within a "hockey puck," or an area that's 15 miles wide and 3,500 feet thick, with the airplane at its center. Extra equipment to receive and display this information in your cockpit, ADS-B In, is available but not required.

As it stands today, the ADS-B system provides air traffic controllers with more complete and accurate information about aircraft, and provides pilots with information that they otherwise would not have at all. To some extent, it is an existence proof that a decentralized system is possible, and it is a step in the right direction. However, it would still require a major revolution at the ATC level to accommodate generally used flying cars.

The best chance that has of happening soon might grow out of a decentralized system for drones. There are now nearly two million drones in the U.S. The FAA tells us about it:

UTM is a separate, but complementary set of services to those provided by the
Air Traffic Management (ATM) system that utilizes industry's ability to supply
services under FAA's regulatory authority that are currently assigned to manned
flight operators/pilots. UTM is designed to support the real-time or near real-
time organization, coordination, and management of primarily low altitude (primarily < 400 ft above ground level [AGL]) multiple BVLOS UAS operations.[230]

(As far as I know, there is no app available to translate this into English.) At the
moment, the only thing that the system supports is automatic notification of airports
of drone activities in the area, but, given the pressure (and an increased willingness to
experiment with drones as opposed to manned aircraft), we may see some advances.
The best chance of seeing some real progress, however, is probably going to come
from the various industry projects to work out distributed drone traffic-control systems. This is where the action is today; as I write, "drone coordination" gets 4.6 million hits on Google.

In the near future, drones will give us an accessible way of discovering what the
issues of coordination are, and of experimenting with solutions. Where the action is
tomorrow will depend strongly on the regulatory environment.

To sum up, the technical problems of building a flying car that could take off from
your driveway and achieve airliner speeds are solvable with today's technology. Such a
car would be too expensive for the average American, but with advancing nanotech
that would change (including the noise). In the meantime, there is a spectrum. It runs
from small planes that need airports and can only do 100 knots to roadable craft, to
helicopters with similar speeds but which can land anywhere, to planes that can do
250, to ones that can do 500 and fly over the weather. Unlike ground cars, where the
extra price of upscale cars mostly shows up as leather seats and sound systems, there
will be a long period where paying more for a flying car will get you more capability.

And at the far end of the process, with full Second Atomic Age technology, you
don't get a flying car. You get a private spaceship.

# 17

# Escape Velocity

*In spite of the opinions of certain narrow-minded people, who would shut up the human race upon this globe, as within some magic circle it must never outstep, we shall one day travel to the moon, the planets, and the stars, with the same facility, rapidity, and certainty as we now make the voyage from Liverpool to New York!*

**—Jules Verne, *From the Earth to the Moon***

How can we futurists, even us nostalgic ones, be so carelessly optimistic about the wonders of technology to come when there has been, in space travel—the most visible and highly hyped of technological applications as of 1962—not only no progress but obvious regression?

As regards space travel, there's a straightforward answer: Project Apollo was a political stunt, albeit a grand and uplifting one. Once the publicity was attained, there was no compelling reason to continue going to the Moon, given the cost of doing so. The nature of political stunts is such that it does you no good to repeat one, even if you can do the same thing better or cheaper.

But there is another, somewhat deeper, answer, which echoes the same basic reasons for our technological stagnation: bureaucratic ossification and incompetence.

On January 28, 1986, I was in an airliner heading home from my grandfather's funeral in Florida. The captain came on the public address system to impart a bit of amusing trivia: The outside air temperature was 90°C below zero, while we were still over Florida. The jet stream had wandered down from higher latitudes, lowering temperatures well below normal, on the ground as well as in the air.

I remember that bit of trivia over three decades later because of its then unknown implications: The same cold wave had stiffened the material of the O-rings in the solid rocket boosters of the space shuttle Challenger, causing a leak of combustion gases that resulted in the breakup of the vehicle. But nobody knew that at the time; it took the Rogers Commission six months to finish its report, with Richard Feynman famously using a glass of ice water to demonstrate the cold-caused lack of flexibility.

Well, not exactly *nobody* knew: It turns out that Allan McDonald, director of the Solid Rocket Motor Project at Morton Thiokol, had warned of the problem and, in fact, had refused to sign off on the launch. He was rewarded for this exercise of responsible foresight by being removed from his position and demoted. Feynman himself had to threaten to resign from the commission and remove his name from the report unless it included his observations of the enormous communications gap between NASA engineers and management.

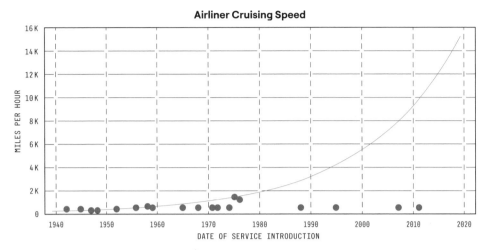

**Figure 46:** Airliner speed curve, with flatline, as seen in Chapter 3, but with the curve extended to date. Low-Earth orbital velocity is about 17,500 miles per hour.

The Dilbertization of NASA, more painfully obvious in later years, was hardly limited to NASA itself. In fact, NASA, especially in the 1960s, probably represents the best-case scenario for such a designed bureaucracy: a collection of the nation's brightest people, with a common goal and vision, committed to a concrete, well-specified task. Remember, "There wasn't actually a sign saying 'Waste Anything but Time,' but it felt like it."

But by the late 1970s, with the end of Apollo and the attempt to lower the cost of space access with the shuttle, NASA simply failed. The shuttle program had the explicit goal of reducing the cost to orbit to $100 per pound, and the lowest they ever got was $10,000 per pound. The shuttle was supposed to launch once a week; it averaged once in 11 weeks. The shuttle blew up, killing everyone on board, about 1.5 percent of the times it was launched. For comparison, if airliners had the same reliability as the shuttle, there would be 1,600 fatal crashes every day.

High on the list of the futures we were promised was the idea of space colonies, as envisioned by Princeton physics professor Gerard O'Neill. It was not just another *Popular Mechanics* article; it was a broadly based movement, the 10,000-member L5 Society, and a substantial body of technical and economic analysis. It foundered almost entirely on the failure of NASA and the Shuttle.

There was something to the Moon landings that was reminiscent of a Lewis and Clark expedition, in that they surveyed the territory, not just the physical but the technological territory of space flight. Notions of what a Moon landing would look like changed drastically from the science fiction of the 1950s to the actuality of the 1960s.

As you can see from the airliner cruising-speed trend curve in Figure 46, we shouldn't have expected to have commercial passenger space travel yet, even if the Great Stagnation hadn't happened. The interesting part of the curve, though, is that it represents the capabilities of the underlying technology—which didn't stop evolving.

**Figure 47:** The fuel pump in the V-2 rocket was rated at 800 horsepower.

We should expect real services to reappear along the same curve when the capabilities match up with some activity that is both useful and economical.

They are about to: The curve predicts that aircraft should get into low-Earth orbital speeds in the coming decade, and orbit is an extremely efficient way to travel. It takes no more energy, total, to get to orbit than a 747 expends dragging its bulk through the atmosphere halfway around the world—and you get to, say, Sydney in under an hour instead of over 20. The beginnings of commercially viable space travel can be seen today. For the first time, private space-launch capabilities are leading government ones in the U.S. If this continues, it is quite possible that we will have a substantial presence in orbit and be back on the Moon before 2030. Fast long-distance travel around the Earth will become increasingly orbital. Once that has developed, there will be the traffic to make orbital space stations economically feasible. It is a possibility.

## Rocket Science

Rockets use a *lot* of fuel. Let me repeat that: Rockets use a ridiculously, gigantically, enormously *huge* amount of fuel. As a kid, I was as much into rockets as my friends were into football. But it took me years of reading, model building, and, ultimately, math to get a decent intuitive understanding of what that really means. My epiphany came when I discovered that the *fuel pump* in the V-2, a small rocket by today's standards, was 800 horsepower. (See Figure 47.)

Why that much fuel? Imagine that you are standing in the middle of the air, 10 miles up. Unfortunately, you will immediately start to fall. You would like to prevent that.

Fortunately, you are standing on a large rock, which weighs as much as you do. Before you and the rock fall much, you jump, pushing off of the rock. It falls faster than it would have, and you stay up.

Unfortunately, you can only jump so hard. You only stay up for a second before you begin to fall again.

Fortunately, you are holding another rock, which also happens to weigh as much as you do.

Unfortunately, that means we were wrong about the first rock. It didn't weigh as much as you do; it weighed as much as you and the second rock, i.e., twice your weight. So you and both rocks weigh four times what you do alone.

Fortunately, your legs are strong enough to jump off the big rock with your load at the same speed, and you stay up just as long.

Unfortunately, to stay up another second you need to have started with an even bigger rock; it has to be the total weight of you and all your other rocks. For each extra second you want to stay up, you have to double the weight of the rocks. For five seconds, 32 times your weight. For 10 seconds, 1,024 times.

And that's why they call them rockets. In a real rocket, the fuel does double duty: It is the reaction mass, doing the job of the rocks, but it also provides the energy with which you do the jumping. In the standard jargon, your specific impulse ($I_{sp}$) is the velocity at which you can jump off the rock, measured in seconds before you need to jump again; in our example it was one. The mass ratio is just the total amount of rocks you bring, in multiples of your own weight.

In real life, the specific impulse of chemical rockets ranges from 300 to 400 seconds.

Virtually every technically literate science fiction writer in the 1950s assumed that we would be using nuclear-powered rockets to explore the solar system. Heinlein's first published novel, and his only movie, featured them in some detail. But probably the best (in the sense of technical accuracy) of the lot was a somewhat obscure writer named Hal Goodwin (who wrote under the pseudonym Blake Savage). Like Isaac Asimov, Goodwin was a practicing scientist as well as a writer; but, while Asimov was a biochemist, Goodwin was the director for atomic test operations for the Federal Civil Defense Administration. He would later move to NASA. His 1952 adventure novel, *Rip Foster Rides the Gray Planet*, was the best account of nuclear spaceships that I read in science fiction. It was the first place I ever heard of thorium, of using nuclear charges to alter the orbits of asteroids, of decontamination, or of radiation sickness.

Probably the best-known design of a nuclear-powered spaceship capable of traveling throughout the solar system is the Discovery One in Clarke and Stanley Kubrick's film *2001: A Space Odyssey*. The design of the ship, with the propulsion unit, including nuclear reactor, at the back, and the crew quarters at the front, separated by a long thin section bedecked with reaction mass tanks, is predicated on separating the crew from the reactor's radiation without an enormous mass of shielding. Frederick Ordway, a NASA engineer who did much of the design work for the fictional craft, wrote:

> The Cavradyne engines are based on the assumption of years of research and development, during the 1980's and 90's, of gaseous core nuclear reactors and high-temperature ionized gases. Theory is presumed to have showed that gaseous uranium 235 could be made critical in a cavity reactor only several feet in diameter if the uranium atomic density were kept high, and if temperatures were maintained at a minimum of 20,000 degrees F. [231]

The real-life NERVA design experimented on in the '60s was, in fact, "limited by the capabilities of solid materials," and only managed to get a specific impulse of about twice that of chemical rockets (800 as opposed to 400 seconds).

Needless to say, the 1980s and '90s did not see "years of research and development of gaseous core nuclear reactors," as Ordway had predicted. Even so, nuclear thermal rockets are the only demonstrated technology capable of navigating the solar system in anything like reasonable times. As I write, things may be starting up again after decades of neglect: The Defense Advanced Research Projects Agency has just awarded a contract to General Atomics to design a small nuclear reactor for space propulsion.

Furthermore, the advances in enabling technologies—such as superconducting magnets, which have caused the recent resurgence in fusion power experiments—could be equally well applied to fusion rocket engines.

No one has demonstrated a nuclear-to-electric-to-ion rocket yet, but it would be the obvious next step. Current prototypes indicate that you could get an $I_{sp}$ of 3,000 seconds or so.

The ultimate reaction rocket, however, is most likely to be a direct fusion-to-jet process, using, for example, the proton + lithium-7 fusion/fission reaction, producing two alphas (helium nuclei) with a total energy of 17 mega electron-volts. Redirecting these magnetically or electrically would give us a specific impulse of roughly two million seconds. With one ton of fuel, your 10-ton family spaceship can accelerate at one G for two days.

This makes all the difference in exploring the solar system. A Hohmann transfer orbit to Mars, the minimum-energy path, which is essentially all that our chemical rockets can manage, takes eight months, and requires a launch window that only happens every 26 months. [232]

This is clearly unworkable as a basis for commerce. However, nuclear rockets, of whatever type, make it possible to fly high-energy low-time trajectories.

The distance between Mars and Earth varies from 50 to 350 gigameters, depending on where they are in their orbits. [233] Accelerating at one G for one day puts you 50 Gm out and traveling at 3.6 Gm per hour. As a point of reference, the Earth and Mars move in their orbits on the order of 0.1 Gm per hour, so, for a typical trip, you accelerate one day, coast a few days, and decelerate one day. Your orbit is pretty close to a straight line. Just watch out where you point that exhaust jet.

# Ad Astra per Koken

*Do I dare*
*Disturb the universe?*

**—T. S. Eliot, "The Love Song of J. Alfred Prufrock"**

A classic 1920s-era Koken barber chair weighs about 250 pounds. It is a solid piece of old-fashioned machinery; many original ones are still in use today. Chairs like that were also something of a luxury item: The Titanic had two of them in its first-class-only barber shop, as did many top ocean liners of the day. But nobody in his right mind would design a *spacecraft* with a barber chair in it!

Well, nobody except Ted Taylor. Taylor was an interesting figure in the history of technology: As a nuclear engineer, he designed both the smallest and the largest fission devices that the United States ever built.[234] Taylor's insistence that there be a barber chair in his spaceship was an in-your-face way of saying that this was a ship, like an ocean liner, and not a dinky little capsule stuck on top of a skyrocket.

Taylor's ship, of course, was Project Orion, the very first study funded by ARPA (later DARPA), the now-famous research arm of the Department of Defense (and the second major project of General Atomics, after the TRIGA reactor). They started studying Orion in earnest in 1959. It was a serious engineering effort, engaging some of the country's top minds (besides Taylor, most famously Freeman Dyson). Some of the design is still classified top secret.

That is because Orion was a spaceship that would be propelled by a string of A-bombs set off behind it. The idea sprang from a nuclear-weapon test in the mid-1950s in which some metal spheres covered in graphite had been set up just 50 feet away from the bomb. They were later discovered to have been blown quite a distance, but they remained essentially unharmed. Stanislaw Ulam, one of the brightest minds at Los Alamos—he was co-inventor of the hydrogen bomb with Edward Teller, and contributed a key element to John von Neumann's theory of self-reproducing machines, among many other achievements—was captivated by the notion that nuclear explosions could be used to drive a vehicle. Taylor caught the bug and was picked to head the project.

The Orion design was 150 feet in diameter and 4,000 tons. Tons. You could afford to put a barber chair in that. You could afford to build it of steel with oceangoing, ship-building techniques. And it was, as far as a team of top technical talent could determine, feasible with the technology of 1960. The scientists seriously expected to be visiting the moons of Saturn by the 1970s. *It was a possibility.*

On the other hand, there is also the far end of the Orion research: the super-Orion propelled by hydrogen fusion bombs instead of small fission ones. Dyson wrote:

> A ship with a million-ton payload could escape from the Earth with the expenditure of about a thousand H-bombs with yields of a few megatons. The fuel cost of such a mission would be about 5 cents per pound of payload at present prices. Each bomb would be surrounded by a thousand tons of inert propellant material, and it would be easy to load this material with boron to such an extent that practically no neutrons escape to the atmosphere. The atmospheric contamination would only arise from tritium and from fission products. Preliminary studies indicate that the tritium contamination from such a series of high-yield explosions would not approach biologically significant levels.[235]

The scientists estimated that the biggest ship they could get off the ground would be about eight million tons. For comparison, a modern high-end cruise ship weighs in at 126,000 tons and carries 3,000 people.[236] An eight-million-ton ship could carry something like 200,000 people, in similar luxury.

One key fact about the Orion technology was that the total atmospheric contamination for a launch was roughly the same no matter what size the ship. That gave the impetus to conceive of ever larger ones.

From the 1951 movie *When Worlds Collide* to the recent Neal Stephenson novel *Seveneves*, one of the staples of science fiction is the end-of-the-world story in which people struggle to launch a remnant of humanity into space to escape the destruction of the planet. Should we be forced to such an extreme, with the mega-Orion we could save the entire human population with 35,000 launches. For comparison, there are 100,000 airliner flights *every day* worldwide.

Orion ultimately foundered on not having a strong mentor in Washington politics, and because of fallout fears. These had led to the atmospheric test bans, which would prohibit any Orion launch.

It is quite possible to engineer bomb/propellent packages that are considerably cleaner than the bombs used in World War II or tested in the 1950s, and, for the really big ships, note that fusion is a lot cleaner than fission, per unit energy released. But it isn't possible to remove radiation completely. If it is true that any amount of radiation, however small, causes some risk multiplied by the number of people exposed, then atmospheric detonations, no matter how remote, will always come with a statistical expected excess cancer rate. So the LNT, the Linear No Threshold hypothesis, makes a huge difference in calculating risk for a radiation-producing enterprise such as the Orion.

Linear hypotheses abound in science, and in plenty of cases it is understood that the phenomenon is nonlinear but a linear approximation is used anyway. One reason is that this makes the calculation enormously easier. Freeman Dyson, who did the fallout-mortality analysis for Project Orion, calculated that you would get 10 excess cancer deaths per launch, explicitly using the linear hypothesis with a figure of one death per 10,000 roentgen equivalent men (rem), equal to 100 sieverts, however distributed.[237]

Of course, what this means, even with the linear hypothesis, is that from the world population of 7.6 billion, 497,040,010 people would die of cancer in the years following an Orion launch instead of the normally expected 497,040,000.[238] But it also means that the calculated dose per person was 0.02 millirem (0.2 microsieverts)—in a world where natural background exposure is about 300 millirem. And the likely protracted no-damage threshold is 50,000 millirem.

Only an extremely doctrinaire adherence to the almost certainly false LNT hypothesis could hold this out as a problem. Yet not only is the hypothesis received wisdom in regulatory circles, but it has made just studying the question virtually impossible. This is yet another social pathology holding us back from technological progress; Orion aside, outer space is a high-radiation environment. We will never inherit the universe until we learn how to live with radiation—and that means studying it honestly.

## The Cold Equations

Allow me to do a bit of non-technological forecasting. As always, the best way to start is to consult history.

The two major wars in North America during the 1700s were the French and Indian War and the American Revolution, which were both proxy wars between England and France. After the end of World War II, most of the wars on Earth were likewise proxies as part of the Cold War, in places that were considered colonies.

If the pattern continues, in the near future I would not be surprised to see a Cold War between the U.S. and China, with Russia and India as contenders. At stake would be control of the oceans and outer space. As a technological forecaster, I can say with some confidence that if we try, both could be very valuable domains and home to significant numbers of people at the end of another century.

Now let us try a thought experiment. How would you react to each of the following:

› The Chinese space force blows up 10 U.S. military spy satellites.
› They blow up a space station with 10 people aboard.

In mere terms of money and national capability, these are essentially comparable. But in political terms they are night and day. It was often remarked during the original Cold War that the American forces deployed in Europe would hardly have slowed the Soviets had the USSR decided to roll the tanks. But although cannon fodder, they were also a trip wire, and their destruction would have given the U.S. the excuse to use nuclear weapons.

It's quite possible that the same dynamics of political logic will contribute to the shape of near-term space exploration and colonization. There may be a lot more people in space than strict economics or scientific necessity would dictate.

But then, that's how the last New World was colonized.

# Cradle

*Earth is the cradle of humanity, but one cannot remain in the cradle forever.*

—**Konstantin Tsiolkovsky**

The young Isaac Asimov, growing up in New York City, imagined as the capital of the Galactic Empire a city that covered an entire planet. In 1,000 years, a relative eye-blink geologically speaking, if human development continues the way it has for the past 10,000, the whole Earth will be like NYC or LA. Compare the beautiful, cool green hills of Earth—say of my erstwhile home in northern Pennsylvania—to the flat, hot, crowded concrete wastelands of the cities. The only animal life left will be zoos, farms, and pets—and with nanotech food production, there won't even be any farms.

This is a consummation devoutly not to be wished. We have to control the population. It's not a question of if, but when. And, of course, how.

The problems are these: Outside of the Four Horsemen of the Apocalypse, the only known way to moderate population growth is wealth. (Note that even in China the birth restrictions are part of an implicit social contract that includes a meteoritic rise in the average income, and even so they are in the process of being phased out.) The technophobic proscriptions of Green fundamentalism are deeply anti-wealth, and thus ultimately self-defeating, as any reduction in individual eco-footprints will be swamped by more footprints.

Lest I seem to be arguing for population control purely from Green motives here, please note that a human-centric ethic produces the same conclusion. Having 10 billion people live short lives in grinding poverty at Level 1 is a horrible alternative to the same 10 billion people living long, healthy lives of opportunity and freedom at Level 4. Or better yet, at Level 5 with flying cars.

So, whether you care about the Earth or the people, you should come to the same conclusion: We need more living room and more energy. We come to the same conclusion again when we think in the long term: The human race shouldn't have all its eggs in the same basket. We must diversify. Fears of a nuclear war wiping out all life on Earth have receded since *On the Beach*, but the possibility remains. As genetic technology begins to experience a growth era, the possibilities of a global pandemic seem more likely than ever.[239] And there are always other possibilities, such as dinosaur-killer asteroids and unknown unknowns.

Carl Sagan is often credited with the quip "All civilizations either become space-faring or extinct." In published works, he said things to that effect, a bit more prosaically:

Most of the billions of species of life that have ever lived are extinct. Extinction is the norm. Survival is the triumphant exception.[240]

Intelligent beings everywhere will have to unify their home worlds politically, leave their planets, and move small nearby worlds around. Their eventual choice, as ours, is spaceflight or extinction.[241]

About 10,000 times as much energy as we actually use today, $10^{17}$ versus $10^{13}$ watts, falls on the Earth as sunlight. But 10 billion times more than that floods through the solar system, available for the taking by anyone who is out there—and energy is the basis of wealth. There are at least tens of thousands of times more usable matter in the asteroids as here—not even counting the other planets! There is plenty of room and resources for hundreds of billions of very wealthy humans without using the Earth at all—we can afford to set Earth aside as a natural preserve or managed park and take our fights (and dangerous experiments) outdoors.

With a large, wealthy, off-world population, we can afford to experiment. We can experiment with different ecologies and governmental arrangements in the various space colonies. We can experiment with the Earth's climate and ecology (which we are already doing, of course) but have the resources to fix it if we screw up, and not place the human race in danger of extinction with every stupid blunder of World Weather Control. We can afford not to fight over the Earth's resources, because they'll be 1 percent of 1 percent of what's available.

The human body is reasonably well adapted for life on the surface of the Earth—at least the part that is land not covered by ice. But even here we need clothing and shelter for large parts of the planet, for large parts of the year. In becoming the most widely distributed animal, we mediate between our bodies and our environment with our inventions. It was always assumed in classic science fiction that the same process would continue in more sophisticated form on the Moon, on Mars, and throughout the solar system. It's fairly easy to envision how nanotech will contribute to that process; in fact, it seems almost certain that nanotech will be the defining level of capability that makes widespread space colonization feasible. In space, we must replace the entire capability of the biosphere to rearrange atoms, and provide the energy necessary to do it.

Less common in science fiction was the extensive modification or complete replacement of the human body itself. One obvious reason is that science fiction needs to provide characters with whom the reader can identify, and this gets harder to do as characters become less human. As long as the modifications are essentially medical, providing radiation resistance or gills, for example, there is no narrative problem. Classic science fiction was full of anti-aging, mechanical education, and similar enhancements. But all the humans looked like humans. It was left for the aliens to look like spiders with a hand on the end of each leg. But wouldn't that be a much more useful form if you were living in free fall?

One of the more amazing things about the High Frontier space-colony movement of the 1970s and '80s was that it actually would have been possible to build Earthlike habitats in orbit with the technology of the 20th century. On the other hand, it would also have been more expensive to do so than the enthusiasts realized. Space colonization is much more likely to succeed when it isn't necessary to re-create a large-scale Earthlike environment for unmodified biological humans to live in.

Given current technology, spacesuits are cumbersome and uncomfortable. We should wish for the vast majority of humans living in the vast majority of the universe that they should be adapted to their environments, that it should feel just as good, not only physically but psychologically, to stand on the Moon or float around an asteroid as it is for us to walk along the seashore on a warm spring day.

One of the major historical watersheds that we can expect from reasonably foreseeable technology in the next half-century or so is the ability to build replacement body parts that actually work better than the original. From peg legs to hooks, this has never been the case before. The coming century will be the century of enhancement. The human of the future will have more and better senses and be stronger and better adaptable to a much wider range of environments.

For the next century, that seems the best description. But in the fullness of time, we will have invented, discovered, developed, and tried new and better forms to wear, as we now wear clothing and carry gadgets. Our bodies may end up being as varied as our machines are now. We will process matter in any form and drink energy from any source. And last but not least, we will have the biosphere atom-rearranging capability built in. The human of the future need not have any ecological footprint at all.

Furthermore, the human society of the future desperately needs a frontier. Without an external challenge, we degenerate into squabbling, self-deceiving Eloi Agonistes. But the solar system is a foeman worthy of our steel; and, after that, the galaxy is even more so. We can pit ourselves against the universe instead of each other for a long, long time.

# 18

# Metropolis

*I look forward to designing a mile-high building.*

—**Dennis Poon, lead engineer, Chengdu Tower**

At the turn of the 21st century, humanity passed a major watershed. For the first time, more than half of us lived in cities. While it may be that our far descendants will stand naked on the asteroids and float in the atmosphere of Jupiter, it seems a not unreasonable prediction to say that, for the coming century at least, the majority of humans will live in environments that are for all practical purposes entirely artificial.

The Burj Khalifa in Dubai opened in 2010 as the world's tallest building, about half a mile high (or about 2,600 noble feet). It broke the record for highest building in 2007, while still under construction, taking the pride of place from Toronto's CN Tower, which had held it since 1975.

Except for being only half as tall, the Burj Khalifa resembles Frank Lloyd Wright's proposed mile-high tower in overall shape—but, of course, the Burj is real. From what I can tell, it could not only house but form the complete social and economic infrastructure for about 5,000 people, if it were designed with that in mind. The actual Burj contains 900 luxury residences and 37 floors of offices. In a shape that wasn't quite so needle-like (e.g., almost any other building), you could have an average of an acre of space per floor, house 25 people per floor with 1,000 square feet per person, and still have plenty left over for public and utility space. That gives *The Jetsons* their 5,000-square-foot apartment, if you count Astro as a person.

Scale the skyscraper up to a mile and you're talking 40,000 people.

Given its elegant shape, the Burj has an incredibly tiny footprint (the foundation slab, not the surrounding plaza) of less than two acres. It seems reasonable to imagine that one could build a mile-high version with a seven-acre footprint. Put only one of these per square mile—it takes up about 1 percent of that area, so the land is left pretty much untouched—and you could house the current population of the Earth in about the area of Montana. Give people flying cars and/or underground high-speed trains to get from one tower to another, and you really could turn the whole Earth into a park not unlike Buck Rogers's America—a *Jetsons* world indeed.

A major engineering load on a tall building is wind, and the load increases with the height of the building. Even though the top of the Eiffel Tower is slender openwork, at the second platform level the structural members on the windward side are in tension, not compression.[242] This is the primary reason that modern skyscrapers are built in interesting shapes, with twists, holes, anything except a simple rectangular parallelepiped. Yes, it does make the building look more interesting, but the main effect is to break up the buffeting vortex street generated by the simple regular shape, and significantly reduce dynamic wind loads.

Wind loading rises alarmingly until about 20,000 feet (four miles), plateaus until 40,000, and decreases thereafter, as the air is much less dense than at sea level. For the same reasons, somewhere between 5,000 and 10,000 feet you begin to have to pressurize the building for people to live in it. (Alternatively, you could increase the oxygen fraction of the interior air to maintain the sea-level partial pressure of 6.28 inches of mercury, 210 millibars. This would work up to about 35,000 feet, i.e., seven miles.)

So, to move toward a world of towers surrounded by parks, you'd have to be able to do a lot of high-tech building, and the people living in the towers would have to be pretty wealthy. Building a world full of mile-high towers with current technology would strain the world's supply of steel and concrete significantly, if it were possible at all. With nanotech, though, it would be pretty straightforward. Some years back, I suggested that a good X Prize for nanotech would be to build a tower 10 miles high. My reasoning was that to do it you'd have to come up with a working manufacturing method to cheaply make the material, probably nanotubes and diamond.

It turns out that that might help, but it may not be necessary. Materials science of the type sometimes called "nanotechnology" has advanced rapidly over the past decades. (And yes, an understanding of materials at the nanoscale is a good part of it.) One thing apparently in the works is an aluminum-steel alloy that might be produced at steel prices but with the specific strength of titanium.[243]

With this, you could build a version of the Eiffel Tower over a mile high with no design modifications besides simple scaling. (The 984-foot original is of wrought iron, with max design stress of about 10 kilopounds per square inch, or ksi, which gives it a safety factor of three.) Optimizing that design would likely make a 10-mile-high structure possible.

It's not so much the strength-to-weight ratio of the new materials you need, however, as something else. I received enlightenment when I talked to a skyscraper structural engineer, who told me, "Steel is light." That was a jarring revelation to someone who had been studying airplane design, but the structural engineer's point of comparison is concrete (for the same strength). However, steel is also springy; a 10-mile steel structure would handle gravity just fine, but would whip around in the wind like an old-fashioned car-radio antenna. You might get a great view from the 520th floor, but you'd need Dramamine for seasickness. Concrete gives you both stiffness and inertia.

The Burj Khalifa used a recent advance: the ability to make concrete of greater strengths than before (and pump it further). Use of diamond or nanotube composites would enable much higher towers. It's a possibility.

When can we expect mile-high, or 10-mile-high, towers? The historic surge in building heights coincided with the Industrial Revolution and the use of iron and steel in building, as exemplified by the Eiffel Tower. The trend line since then has been exponential. It puts the point at which our buildings reach the mile mark in about 2065—just right for *The Jetsons*. However, all the structures in this trend are steel-and-concrete, and so, even though they follow an exponential curve, a shift into nano-manufacturing and materials could easily kick the curve into a different mode. We could even see a major jump, like the Eiffel in 1889, if someone took the new capabilities and set out specifically to build a structure just to be impressive.

A 10-mile tower might have a footprint of a square mile, and it could house 40 million people. Eight such buildings would house the entire 2021 population of the United States, leaving 2,954,833 square miles of land available for organic lavender farms.

This would definitely be an undertaking worthy of 21st-century technology; but the technology—or, perhaps better put, the competence—that we need more is in honest and trustworthy governing, managing, planning, and maintenance. Otherwise, when Detroit falls over onto Windsor, Ontario, it will cause an international incident with who knows what kind of domino effects.

## City Services

*As in the pseudoscience of bloodletting, just so in the pseudoscience of city rebuilding and planning, years of learning and a plethora of subtle and complicated dogma have arisen on a foundation of nonsense.*

—**Jane Jacobs, *The Death and Life of Great American Cities***

Many of the classic SF writers expected automatic package delivery. Philip Francis Nowlan, in the original Buck Rogers novels in 1929, wrote,

Why should he leave his house? Food, wonderful synthetic concoctions of any desired flavor and consistency (and for additional fee conforming to the individual's dietary prescription) came to him through a shaft, from which his tray slid automatically on to a convenient shelf or table.

At will he could tune in a theatrical performance of talking pictures. He could visit and talk with his friends. He breathed the freshest of filtered air right in his own apartment, at any temperature he desired, fragrant with the scent of flowers, the aromatic smell of the pine forests or the salt tang of the sea, as he might prefer. He could "visit" his friends at will, and though his apartment actually might be buried many thousand feet from the outside wall of the city, it was none the less an "outside" one, by virtue of its viewplate walls. There was even a tube system, with trunk, branch and local lines and an automagnetic switching system, by which articles within certain size limits could be dispatched from any apartment to any other one in the city.[244]

Interestingly, Nowlan also predicted radio-controlled flying drones, although only as military equipment and not for package delivery. But it was a common theme in early-to-mid-20th-century science fiction to imagine a city as an engineered transportation infrastructure capable of moving everything from packages to people automatically from place to place.

The value of a city is not to bring people closer together. All other things being equal, I am better off with more space to operate and more distance between me and my snoring neighbor—or the busy pub at 2 a.m., or the sirens telling strangers to get out of the way of an ambulance carrying other strangers.

The value of a city is that living there allows you to reduce the time, the opportunity cost, to get from one place—a home, business, institution, restaurant, recreation—to another. It's pure travel theory. In fact, the only significant variation from the general travel-theory rule that the average person travels 1.1 hours a day, anywhere, is in cities, where they travel a little more of the time. That's consistent with the hypothesis that there are more valuable destinations for a given amount of travel time taken.

I've had a modicum of experience in New York City. My college was on the Erie-Lackawanna commuter rail line, and many of my friends lived in the city. A few years back, I noticed an interesting thing: You could tell, without any reference to external views or other such prompts, whether the cocktail party you were attending was in the city. If the most common single topic of conversation was the weather, you were elsewhere. If it was traffic, you were in New York.

Columnist Megan McArdle, in a retrospective of life in New York, points out how dysfunctional intra-city transportation has become:

> That Saturday night I had three parties to go to, in three parts of the city. I was determined to pack them all in, because when would I see these people again? It took an hour and a half to get to the first one, in Cobble Hill. Inwood and Astoria clearly were not going to happen. As I made that calculation, the incipient panic I'd felt at leaving "all that" vanished, as my city already had. The bits of New York that weren't turning into a shopping mall were instead turning into London, where the cost of real estate pushes the merely affluent people so far to the periphery that it is only really practical to make friends along a single train line.[245]

From Cobble Hill (Brooklyn) to Astoria (Queens) it is 6.5 miles, as the crow flies. From there to Inwood (northern tip of Manhattan) is another seven. McArdle could easily own a car that could cover the total distance in eight minutes, if only there were an open road. In the absence of one, a helicopter could do the same. One would think that, with the density of prospective riders, economies of scale would allow a well-designed and integrated transportation system to do better than private independent vehicles going between ranch houses across the open plains. Instead, it does worse by an order of magnitude.

The leading ideology of city planning, unfortunately, is completely oblivious to this. In fact, the high-density "smart growth" school of thought could reasonably be summed up as motivated by making travel as difficult, rather than as easy, as possible. Like many of our legally latched social pathologies today, this one has the structure of a Baptists and Bootleggers bedfellowship. The Bootleggers are the public transit industry, like the "Portland light-rail Mafia"; the politicians supported by the transit industry's lobbying; and the developers who benefit from subsidies and the power of eminent domain. But who are the Baptists?

They are, to a large extent, the followers of Jane Jacobs, the guru—one might even say patron saint—of densification and walkable neighborhoods that foster thriving communities. In her last book, *Dark Age Ahead*, for example, she tells the horrific story of the 1995 Chicago heat wave, which killed something like 1,000 people, mostly the elderly, of heatstroke and dehydration. It turns out that the deaths were clustered in dysfunctional neighborhoods, where the old folks were found in tightly locked apartments, without friends to help them, too afraid to go out even to seek refuge in an air-conditioned store.

One doesn't need to go to the extremes of a killing heat wave to understand that a dangerous, dysfunctional neighborhood is a lousy place to live. An old-fashioned Elm Street where people sit on their front porches and exchange friendly greetings with pedestrian passersby is an enormously more desirable home. This is, of course, what you get in small towns, and it is a major reason that people are happier there than in cities. Just as with radiation or regulation, a moderate number of other people is beneficial, but an overload is not.

Jacobs is generally read as saying that cars kill communities. This, her acolytes declare, is due to two distinct phenomena. First, putting a highway through the middle of a neighborhood physically separates it, destroys part of it, and makes it unwalkable. Second, once you get into your car you don't interact with people, and the community bonds of familiarity and acquaintance don't form.

This last is why she didn't like suburbs. I can relate to the basic logic of the problem; I lived in suburban New Jersey for 20 years and never met my next-door neighbors. There was nowhere you could go without a car. But those years were far from lonely, non-communitarian ones; in fact, they were probably the most connected years of my life. It is just that the communities I was a part of were communities based on cars, from the restaurants I frequented in nearby towns to the tennis club, to the university where I learned, worked, and taught. All of my friends had cars, too; guests were frequently at my house. Of course, they drove there. The difference was that these communities, like the ones that were just then beginning to spring up online, were voluntary; they were formed by people who sought out the same things and had common interests.

Jane Jacobs did not drive or own a car. It's not surprising that she would have missed the nature of the extended community. If you walk everywhere, you are limited to a literally medieval radius of a mile or so for day-to-day travels. With a car, your radius is 20 times as large, and the area of your community is 400 times bigger. Your friends are just as accessible, for fun and for help, because they have cars, too. The elderly are safer, and feel much safer, in their cars than walking in a bad neighborhood. If the poor folks suffering through Chicago's heat wave had had cars, they could have gotten into them, turned on the air-conditioners, and headed for cooler climes. What about putting a superhighway through the middle of a neighborhood, destroying its pleasant boulevards? Dare I mention flying cars? The traffic on the superhighway neither originates nor terminates in the neighborhood. The cars that would have been on the highway will, in a properly designed system, be invisibly high above our delightful neighborhood.

Knowing all this, then, how would you design a city? Clearly the first, not the last, item of concern is the transportation infrastructure, because the only reason for a city is to reduce the time to get from place to place.

A system designed for the benefit of its customers might look more like this: Lighter rails, built for car-size instead of train-size vehicles. They could be cheaper, more numerous, and run overhead instead of intersecting existing streets. All the vehicles are autonomous car-size pods, privately owned or taken individually on demand, and these go from any station to any other without stopping in between. Having the stations off the main travel tracks has several advantages, notably faster travel times, thanks to the lack of stops, and lower latency, because there are always a few pods waiting at each station. Fancier office or apartment buildings would have their own stations inside. The system as a whole could have a much higher carrying capacity than one with trains and stops, simply due to increased rail utilization. More to the point, it could be nearly as valuable to its riders as having cars. Technologically, this could have been done in the 60s—and indeed it was, in prototype, at places ranging from Disney to RCA. It was too expensive then, but, given how much we've poured down the rathole of worthless light rail since then, we might as well have given it a try.

The greater Dallas-Fort Worth metropolitan area covers nearly 3,000 square miles. If you are on foot, about five of those are available to you. In a car, with present congestion giving you an average speed of about 20 miles per hour, travel theory tells us that you'll have about 380 square miles from which to choose daily activities. A well-designed road system, allowing cars to get their normal effective point-to-point speed of 40 miles per hour, would give you 1,500 square miles, still half the total. A helicopter, or any low-latency VTOL with a travel speed of at least 100 knots, would give you access to essentially all of Dallas–Fort Worth.

I've never designed a city, but I've designed a number of processors and circuit boards. From a circuit designer's point of view, cities are primitive, indeed antediluvian: There's only one level of interconnect (with occasional jumpers). Even the Altair 8800 I built in my dorm room in the 1970s had two levels on its processor board, and today's typically have something like 10.

Even two levels in a city—one for north-south traffic and one for east-west, say—would cut transit times in half. Heinlein's 1938 utopia predicted, or perhaps pre-

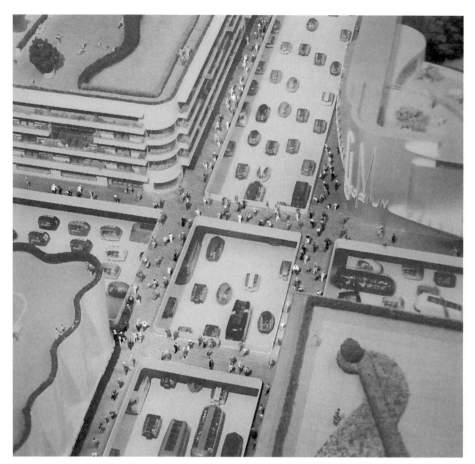

**Figure 48:** Norman Bel Geddes's Futurama in 1939 depicted a city with multilevel, nonintersecting interconnect and separate pedestrian levels. Today's real-world single-level cities remind one of the primitive printed circuits of the '60s: Half the vehicles are standing still at any given time.

scribed, this. Norman Bel Geddes designed Futurama the same way. (See Figure 48.) E. E. "Doc" Smith concurs, by implication: "Northport was not a metropolis, of course; but on the other hand it did not have metropolitan multi-tiered, one-way, non-intersecting streets."[246] A separate level or levels for pedestrians, perhaps with protective awnings like those in Edward Bellamy's 1888 novel *Looking Backward*, could make walking around town a quicker, safer, more pleasant experience, as pedestrians wouldn't have to spend the extra time waiting for traffic, or go the extra distance of crossing roadways.[247]

Given people's revealed preferences, pedestrian levels should probably look more like shopping malls than the useless green space that looks so pretty in planning committees' models (although greenery is valuable in other contexts; see below). On this point, I turn out to be in complete agreement with Jane Jacobs. Indeed, once I started reading her in depth, I was quite surprised just how much I did agree with her.

Lawrence Solomon writes, "She saw government planners as too-often rigid freezers of development, and mocked regulations that prevented, for example, commercial activities in residential districts."[248]

She turns out to be one of those original thinkers whose insights are misunderstood, oversimplified, and misapplied by "followers" who quickly migrate from the Baptist to the Bootlegger side of the equation.

You have to compare the progress in manufacturing or electronics or software over the past half-century with the state of the art in city design and operation to get a sense of just how stagnant the latter has been. Even the recent flurry of utopia-building projects are still basically rows of boxes sitting on the dirt, plus built-in Wi-Fi so that the self-driving cars can talk to each other as they sit in automated traffic jams. Look at a brain, where most of the space is taken up by wiring. Look at a human body, where most of the space is taken up by tubes of one kind or another to move various stuff from one organ to another. Look at a processor chip, a circuit board, a chemical plant, the engine compartment of your car. Designing a city whose transportation infrastructure consists of the flat ground between the boxes is insane.

The crux of the tension between the proponents of densification and the notion of a better world as seen by someone who has to live in it is that there is a perverse incentive. Bureaucrats and politicians wish to force people to interact as much as possible, and indeed to interact in contention, as that increases the opportunities for control and the granting of favors and privileges. This is probably one of the major reasons that our cities have remained flat, one-level no-man's-lands where pedestrians (and beggars and muggers) and traffic at all scales are forced to compete for the same scarce space in the public sphere, while in the private sphere marvels of engineering have leapt 1,000 feet into the sky, providing calm, safe, comfortable environments with free vertical transportation.

There is, of course, the opposing fact that forcing people to interact in a neighborhood tends to make it a better neighborhood. But that is much more appropriate for a village of 250 than a metropolis of 10 million. (If I'm reading Jacobs right, this is another point of agreement I have with her.) An apartment building can, in my experience, be a functioning community, if designed and run properly (and a hellhole if not). The question of whether you get from your building to similar communities and other businesses by "moving ways" or flying cars is orthogonal.

Densification proponents often point to an apparent paradox: Removing a highway that crosses a community often does not increase traffic on the remaining streets, as the kind of hydraulic-flow models used by traffic planners had assumed it would. On average, when a road is closed 20 percent of the traffic it had handled simply vanishes. Traffic is assumed to be a bad thing, so closing (or restricting) roads is seen as beneficial.

Well, duh. If you closed all the roads, traffic would go to zero. If you cut off everybody's right foot and forced them to use crutches, you'd get a lot less pedestrian traffic, too. If you make something more expensive, in time and effort no less than in

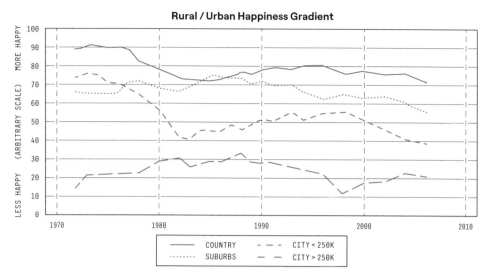

Figure 49:  Self-reported happiness in the General Social Survey.

money, people will demand less of it. It's pure travel theory. The only surprising thing is how the planners' traffic-flow theories can have missed this basic economic fact.

Build more roads and you will get more traffic. The new traffic represents more people now able to do more of the things that they wanted to do but could not before you built the roads. Of course, roads are not free, so you need to balance road cost with value. But no one in their right mind who has any experience with electronic systems design (or who is familiar with nerve pathways in the brain) could imagine that one flat, highly contentious level is optimal for a large, dense, high-traffic system. One might say that 10 levels would be optimal, a wild guess in analogy to processor boards, but even two levels would be a quantum improvement.

The bottom line is that a properly designed city might have 10 levels of traffic interconnect (with parking between travel lanes), a level or two of pedestrian shopping and boulevard sidewalk cafes, and spires towering above that. Travel time from point to point might be a third of what it is in current cities, with the result that the value of living there might be 10 times as great. This, of course, requires a lot of machinery. But what did you think we were going to do with a technology capable of replacing the current capital stock of the U.S. in a week, anyway?

Intelligently managed, with a Henry Adams Curve technology and economy, cities should be machines for facilitating commerce and social interaction. But well-designed cities, with integrated transportation such as moving ways and multilevel streets, are clearly among the high-energy, high-technology futures that we were promised but failed to achieve.[249]

Unfortunately, our current-day cities are among our poorest-performing technologies. Psychological surveys show that people living in the country or small towns are happiest, and those living in big cities are the least happy, with medium-sized cities in between. (See Figure 49.)

Amusingly, people's happiness even correlates with the number of species of birds that they can see and hear.

Not only are people happier in the country, they're healthier, too. You are approximately three times more likely to be murdered if you are in the 50 most populous municipalities in the U.S. than if you're located elsewhere. A 2016 study by Peter James et al. of the Harvard University Chan School Department of Epidemiology found (somewhat to the researchers' surprise) that the benefits of greenery are significant:

> People whose homes are surrounded by the most greenery are 13 per cent less likely to die of cancer. Their risk of dying from respiratory disease also drops by 34 per cent, the biggest ever study into green spaces and health has shown. Overall mortality was 12 per cent less for people who had the most greenery within 250 meters of their homes."[250]

Note the emphasis on homes, not public spaces.

In April 2020, the first wave of the coronavirus pandemic, death rates peaked about 40 percent higher than normal for the U.S. as a whole. In New York City, they peaked at 645 percent higher.

With today's technology, cities and cars are enemies. Every increase in density makes congestion worse and slows cars down. A tower city would be the apotheosis of this trend, making cars useless. On the flip side, the car makes the city less necessary, as the mid-century rise of suburbia demonstrated.

## Little House on the Prairie

*I think it's cathartic to design a dark future, sort of a "glad it didn't happen to me" situation, [but] to design a nice future is a lot more difficult.*

—**Syd Mead**

You could get the same travel-ability as a crowded, highly connected city by building your city as a loose cluster of *Jetsons*-like towers, in which each home and business has a flying-car garage that opened directly to the sky, but which internally contains pedestrian communities of around 250 residents apiece. This keeps you from having to build 10 levels of highway; just use the airspace. What doesn't work with ground cars does work with VTOL flying cars.

We now have the technology to do most shopping online and most delivery with drones, if you have a home with a rooftop landing pad or the equivalent in your backyard. In a readily foreseeable future, nanotech begins finessing even this with ever more capable 3D printers. Similarly, the restaurants, theaters, and workplaces—and friends!—you might want to visit would be as accessible with your VTOL flying car as they would be on foot in a city, but your destination would be perched on a mountainside among the trees, 25 miles away.

Or somewhere. There is no reason for us to be tied down to prescribed little squares drawn on a map. For most of its history, humanity has been nomadic. Our affinity for traveling and seeing new things is deep-seated in our genes.

*The Jetsons*' building need not stand on three improbably spindly legs hundreds of feet above the ground. It can *walk* on three improbably spindly legs hundreds of feet above the ground, like H. G. Wells's Martians. Neither rain nor snow nor sea-level rise nor, if it came to that, glaciations would stay these domiciles from the casual continuance of their ambulatory rounds.

We would need, of course, a major modification to existing law for this to be feasible. But we are going to need something like it once the oceans start filling up with people, anyway.

Current technology is perfectly capable of building seagoing cities; that's what a cruise ship is. Boats or platforms built with current maritime technology are feasible but expensive compared to most (but not all) land-based housing. The places that are more expensive are so because of economic network effects and land-use regulation, for example London, Tokyo, New York, San Francisco, Silicon Valley. What economic advantages might a sea city offer to mitigate this?

At the Seasteading Institute, the original motivating advantage was escape from incompetent government. We can estimate that this might be worth about a doubling of the economic growth rate, compared to the U.S. But that's only going to help if the total cost of doing business can be brought down to within, say, 3 percent of what it is elsewhere. Other factors of production must be found that have advantages hard to find on land.

The obvious big advantage is lots of free space. Your seastead estate can be as big as the private structure you can afford. (And there's no grass to mow.) Here's where technology comes in. As we get closer to true nanotech, the cost of a deep-ocean houseboat should fall, whereas most of the cost of homes on land is due to location.

Mobility is a second big advantage, and this also argues for individual (or smaller-group) structures. Mobility gives you good climate all year round (just follow it), and seasteaders hope that the dynamic geography will ameliorate government lock-in.

Someone living on a cruise-ship-size floating city can walk to everything, but there's not so much there—a city of a few thousand people is quite small, as cities go. A distributed community of seasteads, including larger platforms acting as apartment houses, malls, factories, and so forth, would need a transportation technology equivalent to that of cars and trucks. For heavy freight there are ships, of course, but for fast light freight and personal transportation you'd almost certainly prefer seaplanes. No need to build expensive runways or technologically challenging VTOLs. Seaplanes are a reliable technology that we've had for over 80 years—and we can almost certainly improve on them.

Automatic piloting and cheap power would put the seaplane in the category where improving technology make it feasible as the "family car." And that, in turn, makes seastead communities possible in a much more distributed form, with lower barriers to entry (and exit).

And that's without even thinking about nanotech. With molecular sorting, isotopic separation, and nuclear processes, seawater contains as much energy per gallon as gasoline, mostly as deuterium, lithium, boron, and uranium. It also contains trace amounts of a wealth of other dissolved minerals that a nanofactory would need to augment abundantly available carbon, hydrogen, oxygen, and nitrogen, which would comprise the bulk of its products.

Of course, technology doesn't develop itself—people need to be trying to solve problems, and the problems of seasteading are only beginning to be looked at. Given sufficient interest, however, it seems likely that the technology for dynamic, economically viable ocean communities might be available early in this century, and that completely self-sufficient family-sized units could be possible in later decades. One of the biggest underpredictions that I see for the coming century is the extent to which we begin to colonize the seas.

The ocean liner could be updated to a self-sufficient seagoing city in the 21st century, but it remains basically a Victorian technology. With nanotech, we could do better.

Imagine an aircraft with a 10-mile wingspan. It might be a "flying wing," perhaps shaped like a manta ray. It might have a cord of five miles at the center and a thickness of up to a mile. Internally, there is enough volume for, say, 4,400 square miles of floor area (in 10-foot stories). That is room enough for about 10 million people at 12,000 square feet apiece, room not only for staterooms but for 250 levels of "one-way, non-intersecting" roadways, service cores, 50,000 elevators, and cathedral-like atrium spaces. A 747 weighs about 1,000 pounds per passenger; let's budget five tons per citizen on this airborne city. That gives us a wing loading of about 150 pounds per square foot, same as the 747.

Given the ridiculous wingspan and the virtually infinite Reynolds number, we might get a lift-to-drag ratio of 100; we would need one billion pounds of thrust. With two square miles of thruster area (i.e., a 1,000-foot-wide strip running the entire wingspan), that's 36 pounds per square foot. We will need about 250 million horsepower. Call it 200 gigawatts, with enough left over to keep the lights on. That's just 25 horsepower (19 kilowatts) per person, or just about where we should have been now on the Henry Adams Curve. This might be barely possible with chemical fuels, but one single line of nuclear power plants (every 250 feet) along the wing would suffice to keep Aero City flying indefinitely. That sounds like a lot, but they would occupy only 0.01 percent of the internal volume of the city.

It would strain the world's resources to build Aero City with our current technology, but it would be more or less straightforward with a mature nanotech. Would you want to live there? Only you would know, but it stands as an existence proof that we could live in luxury and leave the entire surface of the Earth untouched.

If you prefer to keep your feet on the ground, remember that an island is just an arrangement of atoms. Dig mile-deep and -wide trenches on the seafloors and use the material to build islands. If you did enough of it, you might even lower the sea level a tiny bit. Mountainous shorelines from California to Hawaii to New Zealand are considered among the most desirable places to live; we could easily fill the Pacific with mountainous islands, barely visible from each other. All easily accessible in our flying cars, of course.

Where are we going to get the energy to do this? The sea floor appears to contain concentrations of uranium ranging from 5 to 500 parts per million, meaning that it contains much more usable energy than if it were solid coal.[251] Consider that a nice added bonus for digging it up in the construction of your island.

It is, as I write, 81° Fahrenheit in Fa'a'a, on the western shore of Tahiti. The expected high for the day is 89. Halfway up the slopes of Mont Orohena, it's 70° or so because of the adiabatic lapse rate of temperature with altitude. Even near the shore, Tahiti is comfortable year-round if you are in the shade, although direct sunlight in summertime is a different matter. Temperatures across the tropical Pacific are similar; the islands are considered a slice of paradise for good reason.

There is easily 10 times the area of the U.S. in the tropical oceans, and all of it has a climate like Hawaii. Swimming in the lagoon, or sunning on the beach, perhaps alternated with sitting on a cool lanai halfway up the mountain looking out on it all, would be a wonderful way to live.

What I would really like to do, though, is float along over the ocean at about 3,000 feet in an open-decked airship, mai tai in hand, watching the endlessly fascinating cloudscapes, the spectacular sunsets, while being wafted from island to island by the trade winds.

Imagine an airship, or rather an airborne village, whose gasbag consisted not of a few big balloons in a fish-shaped cover but a cluster of a billion or so small balloons (10 centimeters would give a gram of lift apiece), randomly spaced and tied together by invisibly thin diamond thread. (Diamond thread the width of a bacterium, i.e., half a micron, should be about right.) The whole business could be made to look, from any reasonable distance, like a naturally occurring cloud.

Below this hangs, perhaps by structures that look like tall trees, shaded from the tropical sun by the "cloud," a village of terraces, courtyards, and esplanades. A center is connected by walkways, while private residences, other villages, and the surface are typically reached by wearable, folding powered wings. Given the nanotech necessary to build the whole cloud city and the wings in the first place, it seems not unlikely that they could be made to look like great white bird wings. It seems much less likely that the inhabitants would be interested in wearing robes or playing harps, however. I wouldn't.

# 19

# Engineers' Dreams

*I'd settle for flying cars. But they need to be real flying cars with antigravity or reactionless thrusters, not ducted-fan kludges.*

**—Glenn "Instapundit" Reynolds**

In *Profiles of the Future*, Arthur C. Clarke listed the invention of a "Space Drive" somewhere in the 2050s. "It is an act of faith among science fiction writers, and an increasing number of people in the astronautics business, that there must be some safer, quieter, cheaper, and generally less messy way of getting to the planets than the rocket."[252] Reactionless thrusters are highly problematical in physics as it is currently understood, but once reaction force is allowed, it is easy to provide it in such a way as to oppose gravity. Tables work, as do balloons, wings, and magnets.

Gravity is not completely understood; if you ask a cosmologist, you will discover that the universe is thought to consist of 5 percent normal matter (i.e., stars and planets), and the rest (i.e., 95 percent of the universe) is thought to consist of "dark energy" and "dark matter." "Dark" in these phrases means that we can't observe them and don't know what they actually are. We just assume they're there because our best theories of gravitation don't match what we actually see the normal matter doing. Is there new physics to be discovered in the field of gravity? Almost certainly. Will this give us a new technology of antigravity? I have no idea.

I am quite confident, however, that there will be a substantial revolution in basic physics: basic as in quantum mechanics. Our theories of gravity seem to be off by a factor of 20, but that is tiny compared to the discrepancy between theory and the observed energy density of the vacuum. Wikipedia puts it:

> Depending on the Planck energy cutoff and other factors, the discrepancy is as high as 120 orders of magnitude, a state of affairs described by physicists as "the largest discrepancy between theory and experiment in all of science" and "the worst theoretical prediction in the history of physics."[253]

As usual, when I try to see a little further, you will find me scrambling onto the shoulders of the nearest giant.

Carver Mead is one of the giants of 20th-century technology. As the Gordon and Betty Moore Professor of Engineering at Caltech, he was a friend, colleague, and collaborator of Feynman. As a friend of Gordon Moore, he was the one who published the technical papers that convinced the electronics world of the phenomenon now known as Moore's Law. (He credits some of the inspiration for this to Feynman's "Plenty of Room at the Bottom" talk.) He invented and built the world's first microwave transistor. When I became a computer processor designer in the 1980s, Mead's book was one of the top essential sources in the field. Beginning with his study of the physics of transistors in the '50s, Mead knows more about how electrons behave than anybody.

So when Mead says that Bohr was wrong and Einstein right, it's worth at least considering the possibility. He says, "It is my firm belief that the last seven decades of the 20th century will be characterized in history as the dark ages of theoretical physics."[254] You'll find 20 times as many Google hits for Einstein as for Bohr today, but in the early-to-mid 20th century Niels Bohr was the bigger name in physics. He was the founder and acknowledged leader of quantum mechanics, and in particular the leading spirit of the Copenhagen interpretation thereof. Einstein was the opposition, claiming that "God doesn't play dice" with the universe. They famously debated the subject as long as they both lived, but in the judgment of the physics community Bohr won all the debates.

The *mathematics* of quantum mechanics has been spectacularly successful, giving unprecedented precision to the statistical prediction of the outcomes of experiments. It is the *interpretation* of quantum mechanics that seems to become completely incoherent. Schrödinger's half-alive, half-dead cat in a box was intended to be a *reductio ad absurdum* argument against it.

Physicists (and philosophers of physics) hold their differing (and contradictory) interpretations with almost religious fanaticism. Why don't they, being scientists, just do experiments to see which is right? The reason is that all the interpretations *give the same answers* for the experiments. They differ outrageously in how they describe the parts that the experimenter cannot know.

In the Copenhagen interpretation, for example, Schrödinger's cat is neither alive nor dead until the box is opened. In the Many-Worlds Interpretation, she is both alive and dead; the universe has split into two possible worlds, and she is alive in one and dead in the other. But for any question that can be answered by measurement or observation, both interpretations give the same answer, and so do all the other interpretations out there.

In 1956, Bohr and John von Neumann visited the Columbia laboratories of Charles Townes. They were there to advise him helpfully that what he was trying to do couldn't possibly work, according to the Copenhagen interpretation. They went away enlightened after Townes showed them a working model. Eight years later, Townes won the Nobel Prize for the invention of the laser.

The central mystery of quantum mechanics is that when energy, say, a photon of light, moves from one point to another, it acts like a wave, refracting and diffracting; but it always comes out all at once in one piece in one place and nowhere else, like a particle. I throw a rubber ducky into the Chesapeake Bay, and a wave spreads out in all directions. (Incidentally, the ducky vanishes.) The wave spreads out all over the world, and then somewhere, say, Copenhagen, with a probability that depends on the height of the wave there and then, someone sees the ducky in the water.

And the wave instantly disappears all over the world.

The Copenhagen interpretation says that this is all there is, all that can possibly be known. The ducky has no physical path between sightings, and, what's more, the wave is all in your mind. It is a little like Skinner and behavioral psychology in the mid-century: a canonical, prescriptive denial of any deeper knowledge. In the case of behaviorism, stimulus and response; in quantum mechanics, "observables" are all you are allowed to know.

In the meantime, the Copenhagen insistence that the electron has to be both a wave and a particle with zero diameter leads to infamous problems, such as the mathematical contortions of renormalization, which Einstein and Schrödinger objected to as resembling "Ptolemaic epicycles,"[255] and the vacuum energy discrepancy.

Carver Mead, not to mention Einstein and Schrödinger and a host of other physicists, such as Fred Hoyle, Jayant Narlikar, Paul Davies, and, most notably, John Cramer[256] (with an assist from Wheeler and Feynman[257]), in the meantime, beg to differ. There is a there there, and Schrödinger's equation, properly interpreted, predicts the quantum phenomena that the Copenhagen interpretation leaves up to magic. The electron is not a point particle predicted by a nonphysical wave; it *is* the wave, just as Schrödinger originally envisioned it. The particle-like phenomena arise from some really beautiful mathematics of wave mechanics, some of which have only recently been worked out.

The reason that these aren't seen in standard practice is that, in true Failure of Nerve fashion, standard practice arbitrarily throws away one of the solutions (the "advanced wave") of the wave equation.

Keeping and using it, instead, leads to a better understanding of a host of coherent quantum phenomena, from lasers to superconductors, to coherent resonant tunneling rectennas, to Bose-Einstein condensates. There remains an enormous amount of work to be done, and we cannot say with certainty exactly what a "Copernican" interpretation of quantum mechanics will entail. Remember that Ptolemaic epicycles gave more accurate predictions of the planets for a century after Copernicus, until Kepler and Newton. But, ultimately, a new perspective on quantum mechanics will be more intuitively satisfying, and will shed light on a host of new mechanisms. That should be incentive enough for progress.

## Pier Into Space

In the meantime, let's see what just plain old, easily foreseeable Second Atomic Age technology could do—could be doing now if Feynman's path had been taken—to make space travel more achievable and affordable.

It's widely understood how lighter, stronger structures can make rockets more efficient, but that's of limited use. The rocket equation is still a huge stumbling block.

One way around the rocket equation is not to use rockets at all. With nanotech we could easily build a space pier.[258]

The idea of a space pier is just the same as for a pier for ordinary boats: It bridges the gap between the land and the regime where the ships operate. In the case of water ships, that's water of a certain depth; in the case of spaceships, it's vacuum and orbital velocity.

Build a structure 100 kilometers tall and 300 kilometers long. Due to the curvature of the Earth, the center is bowed nearly two kilometers away from a straight line; the accelerator track on the top is nearly five kilometers longer than the base (even though both end towers are locally vertical). The tower extends through the stratosphere (50 kilometers) and mesosphere (90 kilometers), extending into the ionosphere. Put a linear induction (or other electromagnetic) motor along the top. An elevator goes straight up 100 kilometers to the top. Payloads are then sent horizontally into orbit with an acceleration of only 10 G's (which appropriately cushioned humans can endure for the 80 seconds required).

Like any other launch-track-based scheme, this has the huge advantage that you're only accelerating the payload to orbital velocity, instead of accelerating 10 or 20 times the payload's mass of fuel and oxidizer.

With Second Atomic Age technology, a tower 100 kilometers high is easy. Flawless diamond, with a compressive strength of 50 gigapascales (GPa), does not even need a taper at all for a 100-kilometer tower; a 100-kilometer column of diamond weighs 3.5 billion newtons per square meter but can support 50 billion. Even commercially available polycrystalline synthetic diamond with advertised strengths of five GPa would work. Of course, in practice, columns would be tapered so as not to waste material, and the base of the tower would be broadened to account for transverse forces, such as the jet stream. Only the bottom 15 kilometers (i.e., 15 percent) of the tower lies in the troposphere and would have to take weather into account.

The electromagnetic accelerator along the top might be fairly heavy. In many designs, coils have iron cores; NASA's Marshall prototype weighs 100 pounds per foot. If we allow a ton per meter, the total weight of the accelerator is 300,000 tons. For comparison, the Golden Gate Bridge is about 800,000 tons. However, most of the weight is (relatively cheap) iron. In fact, even if we had a lighter material available, we might build the track heavy anyway, for stability. Note that this entire weight, if it were concentrated in one place, could be supported by a column of currently available polycrystalline diamond less than 80 centimeters on a side. The overall structure could be openwork, like a radio tower, and might have approximately 60 footprints 10 kilometers apart on the ground—if we set aside a hectare for each foot, they only occupy 0.02 percent of the land under the tower. The footprint foundations would each bear the weight of a small office building, no great technical challenge.

Another connection to the Feynman Path, beyond simply needing nanotech to start with, is that the construction of the tower would be facilitated by tele-operated scale-shifting robots, this time to larger scales, instead of having people in spacesuits crawling around 60 miles up in essentially a vacuum.

The reason for the 60-mile (100-kilometer) height, sometimes called the Kármán line, is that it's just the edge of space. The air is about a million times thinner than on the ground, but there's still enough of it that nothing orbits at that altitude for long.

The tower would be maybe a million tons of material in a structure 300 kilometers long. A typical superhighway that long involves 15 million tons of material and costs on the order of one to five billion dollars. Assuming the land under the tower does not lose too much value (and it might well gain it, especially near the embarkation port), the area occupied by the footprints is trivial. The major obstacle to construction is likely to be legal hassles.

The energy cost of sending a 10-ton payload to the top of the tower by elevator is about 10 gigajoules (2,778 kilowatts, or $138.89 worth of electricity at five cents per kilowatt-hour), which amounts to 1.4 cents per kilogram. At an express elevator speed of 50 meters per second (108 miles per hour), it takes over half an hour to get to the top. It is not clear that there is a compelling reason to go faster, since, unlike rockets, elevators can be made efficient at as slow a climb rate as desired.

Inert freight would presumably be packed into projectiles on the ground, but passengers will want to wait until getting to the top. (Nice view! The horizon is over 1,000 kilometers away. We should probably put a revolving restaurant up there.) At 10 G's, launch acceleration, passengers will be best off ensconced in form-fitting, fluid-filled sarcophagi. Even so, the launch will not be particularly pleasant. (The restaurant will excel at light fare.)

The vehicle is then accelerated at 10 G along the top of the tower for 80 seconds, to eight kilometers per second. The track is level, and with no lateral acceleration the vehicle is in an orbit with a 100-kilometer perigee and 500-kilometer apogee. (The numbers include an assist from Earth rotation.) The energy required is 300 gigajoules ($4,166 worth of electricity, today's prices, again for a 10-ton payload, i.e., 42 cents per kilogram). Call it a dollar with overhead and amortized construction. But given Second Atomic Age technology, both in energy and the ability to re-create the entire physical capital of the U.S. in a week, that would drop to pennies.

The power (as opposed to energy) requirements average 3,750 megawatts for the 80 seconds. The tower's power draw increases linearly from 0 to 7,500 megawatts during the 80 seconds of a launch; a typical suburb on the same land might draw a peak load of 750 megawatts. Local short-term energy storage, in a form conducive to rapid drawdown, will be necessary for load averaging. Energy storage (and release) requirements are constant per track length and amount to one megajoule per meter. (One MJ is the energy in 15 pounds of lead-acid batteries, an ounce of butter, or a speck of uranium the size of a grain of salt.) It might even be possible to store energy in the magnetic fields of the accelerator coils. Drawing at a typical power station's production of 1,000 megawatts, it would take five minutes to recharge the tower.

Your payload needs to be a spacecraft capable of some 330 meters per second delta-V to circularize its orbit. The best strategy is to inject it into an orbit whose energy is the same as the desired circular one, and do a correction at the point the orbits cross, which only changes direction. This is not a Hohmann transfer orbit, which optimizes total delta-V. With the launch tower, delta-V at correction is considerably more expensive than at launch, so we minimize it preferentially. Even so, given typical chemical specific impulses, propellant for correction is some 10–15 percent of gross vehicle weight. (If it were a high-$I_{sp}$ direct fission rocket, of course, it would be trivial.)

In the early days of flying, there was a saying: Zeppelins breed like elephants, but airplanes breed like rabbits. A realistic guess is that, ultimately, we will build space piers, but only after the volume of traffic is already such as to make it economically viable.

Your 10-ton Second Atomic Age vehicle is already a car, a boat, and an airplane. How much extra would it take to make it a spaceship?

From the point of view of the passengers, a 1-G thrust would be easy to take; this gets you to orbital velocity in about 13 minutes. You'll travel 3,000 kilometers in the process; for each kilometer, you will use 100 megajoules of energy. If we can get, say, nitrogen fusion to work, about a cubic foot of air is all the fuel you need. Otherwise, for the total of 300 gigajoules, call it 50 cents' worth of uranium, if you can get around the rocket equation.

The best way to do that, as long as you're still in the atmosphere, is probably electric jets (ionic thrusters, not RC-model fans). They can have high power-to-thrust efficiency, and there's no obvious reason why they wouldn't work at hypersonic speeds.[259] No need to try and support supersonic combustion or put the air through a thermodynamic cycle, as in even a simple nuclear-powered ramjet.

All you need is a power source that peaks at 800 megawatts. Failing that, you have to reduce thrust as you get near orbital speed. And you are going to have to have a rocket to maneuver once you get to space, anyway.

Of course, you could always stay home.

# World Weather Control

Clarke's *Against the Fall of Night*, one of the great classics of golden-age SF, is set a billion years in the future, when the Sun will have expanded to the point of boiling away all the Earth's oceans, leaving only dry, salt-encrusted seabeds on a lifeless planet.

On the other hand, Fritz Leiber's marvelous short story "A Pail of Air" has the Earth being flipped out of orbit into cold interstellar space by a passing dark star, causing all the air to freeze into a deep layer of snow.

In Neal Stephenson's recent *Seveneves*, continuing the classic tradition, the Moon inexplicably explodes and the rain of the resulting meteors boils and burns the Earth's surface for 5,000 years. The same thing had happened to the planet Metaluna in the 1950s SF movie *This Island Earth*.

In the novel and iconic George Pal movie *When Worlds Collide*, a rogue planet actually hits the Earth. Assuming a closing velocity of 30 miles per second (the planet is supposed to have fallen in toward the Sun from interstellar space), the entire Earth would be briefly converted into a quarter-million-degree plasma.

Now *that's* global warming.

Anthropogenically induced climate change today is a favorite hobbyhorse of ergophobic activists, Baptists and Bootleggers both. It is worth pointing out, yet again, that climate change was *not* the reason for the Henry Adams Curve flatline in the 1970s; the notion didn't get significant traction until the late '80s. Furthermore, the horror-story bias of the media has served to grossly misinform the general public, compared with what scientists and economists actually understand.

It would require much more space than we can spare to address the subject of economic trade-offs in fossil fuels versus climate.[260] Luckily, we can simply ignore the issue. Second Atomic Age technology could easily power civilization without any $CO_2$ emissions at all. We should have been building, experimenting, and innovating with it for the past 50 years. Furthermore, to really reclaim our birthright and an optimistic future, we must get back on the Henry Adams Curve. Burning fossil fuels in Earth's atmosphere isn't going to support that. Burning fossil fuels to explore and develop the solar system is downright laughable. At what rate we should switch over is a matter of debate, but switch we must.

In fact, the major problem in the Second Atomic Age era is most likely to be too little $CO_2$ in the air rather than too much. Go look at a cornfield: As I write, it is still weeks until the Fourth of July, but in these parts the corn isn't just knee-high, it's more than head-high. All of that plant material was created by molecular machines, powered by energy from the Sun, out of carbon from the $CO_2$ in the air. In still air on a clear, sunny day, a cornfield depletes all the $CO_2$ in the ambient air in five minutes flat.

Your pocket synthesizer, the Second Atomic Age equivalent of the iPhone, will build you anything—getting the carbon the same way.

Just by way of context, too little $CO_2$ in the air is a lot more dangerous than too much. Current levels (around 400 parts per million) are lower than optimal for green plants; commercial greenhouses operate at 1,000 ppm. If we cut $CO_2$ in half, we would be in serious danger of starving all the green plants of Earth.

So, when everyone starts carrying a pocket iPrinter capable of conjuring up anything from snacks to items of clothing out of thin air, we may have to reopen the coal mines and pump out $CO_2$ just to keep our ecosystem from collapsing.

Current efforts to regulate the Earth's climate cost the world economy something on the order of a trillion dollars a year, not counting the externalities of energy poverty and so forth. What's more, they are not succeeding: Global $CO_2$ emissions are rising faster now than before the Kyoto Protocols were adopted.

Suppose we instead used that trillion—roughly 50 times NASA's budget—to build a robust orbital infrastructure and simply put up some sunshades. The total area needed to negate the enhanced greenhouse effect would be a two-mile-wide ribbon around the equator. Then not only would you have cooled the climate, you'd have a robust orbital infrastructure, a vastly more valuable thing. For example, if it was a solar power satellite instead of merely a shade, it would generate more power than the human race currently uses. More to the point, you would have decoupled the Earth's temperature from the atmosphere's $CO_2$ content. There is no reason to believe that the optimal levels of the two are necessarily in lockstep. With Second Atomic Age technology, it would be essentially trivial to adjust the $CO_2$ level in the atmosphere to whatever we might want. It is only a rearrangement of atoms.

On the other hand, you might not have to launch your shade into space at all. Construct a small aerostat—a hydrogen balloon—at a size somewhere on the order of a centimeter in diameter. It has a very thin shell of diamond, maybe just a nanometer thick. It's round, and it has inside it an equatorial plane that is a mirror (possibly extending outside, giving it the shape of the planet Saturn). If you squashed it flat, you would have a disc only a few nanometers thick. Although you could build such a balloon out of materials that we currently use to build balloons, it would not be economical for our purposes. Given that we can build these aerostats so that the total amount of material in one is actually very, very small, we can inflate them with hydrogen in such a way that they will float at an altitude of 20 miles or so—well into the stratosphere and above the weather and the jet streams.

Each aerostat contains a mirror and a control unit consisting of a radio receiver, computer, and GPS receiver. The device has just barely enough power and fans or other actuators to tilt itself to a preferred orientation. That's all it does—listens for commands on the radio, and tilts to an angle that is a function of its latitude and longitude. It's not a complicated machine.

Now make enough of these to cover the entire globe. For the centimeter size, you'd need about five quintillion. This is why nanotechnology makes a big difference. If you tried to cover the Earth with something the total thickness of even a current-day party balloon, let's say about 100 microns, you'd need on the order of 100 billion tons of material—but with the nano-engineered design, just a few nanometers thick, you'd only need about 10 million tons. To compare that with the scope of current-day construction, 10 million tons is roughly the amount of material that is used to make a hundred miles of freeway. This is an amount of material that current-day technology, much less nanotech, can handle straightforwardly.

That is the Weather Machine. We have aerostats that float 20 miles up. They have GPS and controllers and can turn themselves. That's all there is to it. What could you do with a machine like this?

The machine is essentially a programmable greenhouse gas.

If you set the mirrors facing the Sun, they reflect all the sunlight back. If you set them sideways, they allow the sunlight to come through, and similarly for the upwelling long-wave radiation carrying the energy back out into space. It may help to remember that the amount of energy radiated away by the Earth is the same, to within

**Kardashev Type I Civilization**

| TYPE | POWER | ORIGINAL DEFINITION |
|------|-------|---------------------|
| I | 17 | Sunlight hitting planet |
| II | 27 | Total power of Sun |
| III | 37 | Total power of galaxy |

Figure 50: Kardashev's scale. Power numbers are log watts.

tenths of a percent, as the energy coming in from the Sun. We can cool off by shading the Sun. We can warm up by reflecting back our own emitted infrared.

For mere climate control, we'd only need to build a few tenths of a percent of a full Weather Machine, enough for just a few watts per square meter. The controls on the individual aerostats can be very simple. When set to let sunlight through, for example, the mirrors can be several degrees away from edge-on to the Sun, and the effect would be a scattering of light (visible perhaps as a slight haziness) but no significant reduction in total insolation.

While *The Jetsons* was in its first-season reruns, Russian astronomer Nikolai Kardashev proposed a three-level scale for advanced civilizations. It is worth trying to wrap your mind around the scope of his vision; he saw civilizations on planets, throughout solar systems, throughout galaxies. And he erected a standard with which to compare them. (See Figure 50.)

By his standard, the ratio between his galactic civilization and the ergophobic Eloi only-one-Earth vision that we are being forced to swallow is roughly the same as that between our current global human civilization and a single bacterium.

A Kardashev Type I civilization is one that controls all the energy available on a single planet, which is usually estimated as the amount of sunlight hitting it. A Weather Machine would do that—humanity's total current energy use (strictly speaking, power use), at $2 \times 10^{13}$ watts, is a negligible fraction of the over $10^{17}$ watts that the machine would control.

A Weather Machine could probably double global GDP, simply by regional climate control. The economic benefits would be enormous. There is a huge amount of value in being able to control the weather—excuse me, the climate. If you could tailor it, you could make land in lots of places on the Earth, such as Northern Canada and Russia, as valuable as in California. Weather control is something that people have always wanted to do: 1950s science fiction is replete with it—not to mention the mythic stories of most major religions. Therefore, once it becomes possible to do, we probably will.

The better control we have over our aerostats, the more we can do. Controlling insolation on the scale of tens or hundreds of miles would probably give us the ability to affect daily weather patterns, as well as climatic averages. A hurricane is a self-organized heat engine running on sea-surface heat. Shade the areas where you don't want it to go, and you can warm a path to send the storm harmlessly off into mid-ocean.

In 2029, the asteroid Apophis is going to make a close approach to the Earth, within the orbit of the Moon. It will be something of a butterfly-effect incident; it is fairly difficult to predict exactly what the asteroid's orbit will be after that close encounter. Apophis will return in 2036, and we can't say for sure whether it will strike the Earth or not. In 2029, if a highly controllable Weather Machine has been built, as the asteroid comes tumbling past we could focus a few petawatts of sunlight on it and give it a kick. A small kick in 2029 will have a huge result in 2036, almost certainly enough to prevent a strike.

Take the same aerostat, but inside put an aerogel composed of electronically switchable optical-frequency antennas. We can now tune the aerostat to be an absorber or transmitter of radiation in any desired frequency, in any desired direction (and if we're really good, with any desired phase). It's all solid state, with no need to control the aerostat's physical attitude. Once we have that, the Weather Machine essentially becomes an enormous directional video screen, or, with phase control, a hologram. Call it Weather Machine Mark II.

Astronomers would have hated Weather Machine Mark I, because it acted like a layer of permanent haze in the atmosphere. But they love Mark II, because it turns the entire Earth into a telescope with an aperture of 8,000 miles. Mark I could zap Apophis as it flew by inside the Moon's orbit. Mark II is a little more impressive. At the closest approach of Mars, with a transmitting aperture of 10,000-kilometer diameter, using violet light for the beam, you could focus a petawatt beam on a 2.7-millimeter spot on Phobos. A petawatt is about a quarter megaton per second; 2.7 mm is about a tenth of an inch. In other words, you could blow Phobos up, write your name on it at about Sharpie-handwriting size, or ablate the surface in a controlled way, creating reaction jets and sending it scooting around Mars in curlicues like a bumper car.

Mark II, with the ability to shift frequencies and directions independently, is powered at night. Mark I could cool the Earth by shading the sunlight on the daylit side, or warm it by reflecting back the infrared that pours into the night sky. The total power going in and out is very close to the same; incoming power at the surface is roughly a kilowatt per square meter. That is, per square meter of surface directly facing the Sun. You only get the full kilowatt at noon near the equator. The whole surface of the Earth is four times as large as its Sun-facing profile, so the outgoing power is about 250 watts per square meter, but everywhere, all the time.[261]

Mark II absorbs the outgoing infrared and uses it. Thus, there's plenty of power available for the night side to do street-lighting, or show ads in the sky, or whatever you'd like. Remember, it's a hologram, so it can have a completely different effect for each spot on the surface: My night sky can be a giant telescope, and my neighbor's can be a giant video game.

Somewhat more usefully, Weather Machine Mark II could simply focus transmitted power to any flying car, not to mention any ship, train, ground car, or building. For direct conversion to electricity, it should probably be in the microwave atmospheric window at 15 gigahertz (GHz) or less (with a wavelength of 20 millimeters or longer).

You would need less than a one-kilometer patch of Weather Machine to focus on a one-meter rectenna at 15 GHz; that much sky would have 250 megawatts available at midnight.

I'm fairly certain that a Weather Machine, or something like it, will be built sometime this century. The apparatus, particularly the Mark I form, seems straightforwardly within the capabilities of molecular manufacturing. There are plenty of people worried about things like climate change or asteroid impact. It could have an enormous economic value. All of these indicate that it should be built, but the most pressing and cogent reason that it actually will be is likely military. A Weather Machine would be a very potent weapon, even the Mark I version. Mark II, as noted, could shoot down the moons of Mars.

Even without direct attacks, whoever controls the weather on Earth will be pretty much in charge. Anyone who objects and starts rattling their sabers gets 20 years of no summer or spring. Good luck with those farms. For that reason alone, given the technological capability of doing it, it will likely be done.

If a government managed to build approximately 5 percent of a Mark I Weather Machine, it would have a dead man's switch: If someone came and blew you up and you quit sending out the control signals, all of the aerostats would default into snowball-Earth mode. It would be a doomsday device. This is troubling.

So once somebody gets 5 percent of one built, you're stuck listening to them. You had better start building your own first, or at least simultaneously. In fact, it seems reasonable to imagine that, by later in the century, there are going to be several competing clouds of these things around. We can all hope that they won't end up physically competing with each other, and that the people in charge of them will come to some negotiation. That's going to be all the more reason for someone wanting to be in the game. You have three quintillion balloons up, and I have one quintillion, and this guy over there has two quintillion, which means we get that many votes in the weather-control world government.

The ultimate implications of a Weather Machine are mind-boggling. I can't even come close to seeing all of them, but I'm fairly sure that it's possible and that it will happen. It is a possibility, and one particularly worth thinking about.

## Kardashev vs. the Flying Saucers

One of the more enduring tropes of science fiction, beginning with Wells's *War of the Worlds*, is the alien invasion. Luckily, it now appears that the solar system is not teeming with intellects vast, cool, and unsympathetic who regard this Earth with envious eyes. On the other hand, we have no such guarantee about the rest of the galaxy. We haven't found alien civilizations yet, but among the theories of why we haven't are such hypotheses as someone came through and wiped them all out, or they are keeping their heads down against just such a possibility.

If we can make Weather Machine fabric at less than about a gram per square meter, we can put as much of it as we want anywhere in space, beyond orbit, supported against the Sun's gravity by the Sun's radiation. We say "anywhere" because, to a first approximation, the radiation and gravity scale the same way with distance from the Sun.

This means that we aren't confined to Weather Machines floating in the skies of Earth; we can put an Earth-size patch of Weather Machines anywhere. There's room for about two billion of them at the distance of the Earth from the Sun. If we cover the surface of a sphere around the Sun, we have what's called a Dyson sphere, though Freeman Dyson himself credits the notion to British science fiction writer Olaf Stapledon.

If we did build such a gadget—let's call it a Weather Machine Dyson Sphere, or a WMDS for short—we could make it of heavier material than the simple light-sail calculation implies. In theory, if the inside were perfectly reflective, you wouldn't need the continuing sunlight at all: Light ricocheting around inside would suffice. It would be a balloon inflated with pure light. In practice, you would have losses, which the Sun would make up; the rest of the sunlight would be available as usable power.

Of course, surrounding the Sun with a reflector would make it, the Sun, run hot, which might be useful if we decided that we needed more power. On the other hand, we might be well advised to make the sphere smaller than Earth's orbit. We would, of course, want to program a transparent spot to follow the Earth around.

Using a WMDS half the width of Earth's orbit as a telescope, you could resolve features about six inches wide at the distance of Alpha Centauri. Conversely, you could pour a quarter of the Sun's power output (i.e., 100 trillion terawatts) into a six-inch spot that far away, making outer space safe for democracy.[262] Just remember that what you see was actually there four years ago, and your beam won't get there for another four.

These gadgets, particularly the WMDS, seem like big, powerful machines. But it is important to remember just how small the power density of such things would be in Second Atomic Age terms. Given the analyses of Projects Orion and PACER, one can imagine an enormous piston engine where the cylinder explosions are H-bombs. One probably would not couple the piston(s) to a Brobdingnagian crankshaft, however; a free-piston design that generated electric power directly by shoving magnets past coils would be more reasonable. This would be quite a powerful machine. Just one cylinder using one-megaton charges at one stroke every four minutes would produce 15 terawatts, otherwise known as the total current human energy usage. It could power a space pier or Aero City without even registering the power drain. Replacing each existing power plant in the U.S. with one of these engines makes you a Kardashev Type I civilization at $10^{17}$ watts, equaling all the sunlight that falls on the Earth. And this is *First* Atomic Age technology: We could have built this in the 20th century. Your city, or city-sized spaceship, can have several such engines, each a few thousand feet in size.

Here's another way of defining the Kardashev scale: How much matter does your civilization convert to energy, say in kilograms per second? By a nice coincidence, the number for a Type I civilization is one kilogram. The number for Type II, which would correspond to the amount that the Sun converts, is close enough to 10 billion, the ratio between types in the original scale. (There are 100 billion stars in the galaxy, but many of them are red dwarfs, so 10 billion works for the Type II:III ratio as well.)

To put this in perspective, right now the average human uses about 1,400 watts; the average American uses 10,000 watts or 10 kilowatts. In a Kardashev Type I civilization with the Earth's current population, the average person would use 14 million watts. If we split the difference, say 100 times the people using 100 times the power, each of the 700 billion people uses 140 kilowatts. That matches a civilization with flying cars and colonies throughout the solar system. But it's still only the amount of energy that hits the Earth! There's 10 billion times more, flooding through the void, going to waste. A WMDS could capture all of it, beaming power to every dwelling and machine in cis-Plutonian space. That was, after all, the point of Dyson's concept.

# 20

# Rocket to the Renaissance

*The rapid Progress true Science now makes, occasions my Regretting sometimes that I was born so soon. It is impossible to imagine the Height to which may be carried in a 1000 Years the Power of Man over Matter. We may perhaps learn to deprive large Masses of their Gravity & give them absolute Levity, for the sake of easy Transport. Agriculture may diminish its Labour & double its Produce. All Diseases may by sure means be prevented or cured, not excepting even that of Old Age, and our Lives lengthened at pleasure even beyond the antediluvian Standard. O that moral Science were in as fair a Way of Improvement, that Men would cease to be Wolves to one another, and the human Beings would at length learn what they now improperly call Humanity.*

**—Benjamin Franklin**[263]

If your hobby is traveling to wine-producing areas and going on tasting tours, you may have noticed a trend: Wine grapes are being grown in drier and drier areas, eastern Washington State and West Texas, for example. The reason is that in an area where the natural weather is perfect, on the average, for grape growing, the conditions still vary around that happy medium. However, in, say, West Texas, where it is usually drier than optimum, you can irrigate up to the right amount precisely.

One of the major technological revolutions occurring this century spells the approaching end of humanity's first major world-changing technology, agriculture, after its hundred-century run. The revolution looks like this: Build a big, warehouse-size building. Inside, it looks something like a Walmart, but with the shelves going all the way up to the ceiling. On each shelf grows lettuce, lit by special LEDs on the shelf above. The LEDs emit only the frequencies used by chlorophyll, so they are an apparently whimsical purple. The air is moist and warm, and has a significantly higher fraction of $CO_2$ than natural air: ideal for growing plants. This is not a greenhouse, however, because it does not depend upon the Sun. Supplied with power, it keeps the plants growing 24 hours a day, 365 days a year. The water is recycled, the air is filtered and enriched, and the plants do not need pesticides, because pests can't get to them. Add all the advantages together and you get something like 300 times as much lettuce per square foot of ground as the pre-industrial mule-and-plow dirt farmer could produce.

All you need to have fresh local strawberries in Siberia or avocados in Antarctica is power.

That is today. The near-term trajectory is pretty obvious—more and different kinds of plants, plants bred or genetically engineered for the environment: stalkless corn, fruit bushes. *Vinifera* (wine grapes) as currently grown are already close to this form. There are substantive efforts to produce lab-grown meat. Early efforts in 2013 produced meat at $325,000 per pound; five years later, it was down to $363. Prospects for getting that down to a competitive level are good enough that they are attracting substantial investments from agri-giants, such as Tyson Foods. As I write in 2021, Singapore has just approved cultured bioreactor-grown chicken meat, which is now on sale to the public there.

Something like 80 percent of the land under cultivation in the U.S. produces feed, mostly corn, soybeans, and hay. A much greater chunk of land goes to direct grazing. Right now, all of that land serves primarily as a very inefficient solar collector, gathering the energy necessary to drive the rearrangement of atoms from $CO_2$ back into hamburgers. Second Atomic Age technology not only short-circuits the laborious biological process, but it also supplies the energy. We get half the country back to live on or use as we please. By the end of the century, agriculture, the defining human activity for 10,000 years, will end, and a new human era, based on limitless power and physical capability, will begin.

## Then What?

Here is the choice that's facing us in the coming century: Will we, as a society, pick a comfortable, static level of existence, requiring a modest amount of production that robots could easily supply? Or shall we put a flying car in every garage, usher in the next Atomic Age, and inherit the stars?

Go through door number one and humans could essentially retire, quite comfortably by 20th-century standards. Which means, in practice, that they will spend a vast amount of energy on make-work, self-deception, and, as Benjamin Franklin put it, being Wolves to each other.

The theory of evolution as understood in the 1890s was not "a mass of errors," but it did have some significant gaps. H. G. Wells wrote *The Time Machine* before game theory was developed, and before computers allowed us to do extended simulations of society using its mathematics. Game theory, computer simulations, and history suggest that he missed the mark. Even safe, comfortable humans have not degenerated into mild, harmless Eloi; in the sheep's clothing of virtue signaling hide the Wolves.

Forty years on from *The Time Machine*, Wells had a firmer grasp on this, based on both personal experience and keen observation. The Do-Nots of *Things to Come* end the movie swarming the Moon launch in true peasants-with-pitchforks fashion. Fifty years from that, evolutionary game theory was able to show us why: The evolutionary pressures on what we consider moral behavior arise only in non-zero-sum interactions. The zero-sum society is a recipe for evil. A world where a productive cadre—or robots, or anything exogenous to the mass of people—did all the work would not be a place where I would want to live.

The opposite vision is of a dynamic society. Increased productivity, instead of allowing fewer people to produce the same fixed output, can allow everyone to do more and more. This is a world of makers instead of takers. This is the world in which everyone can afford a million-dollar flying car and vacation around the rings of Saturn. This is the world where we build cities on the sea and colonies in space. This is the world in which humanity is able to preserve the Earth as a park and take our fights outdoors. This is a place where I would want to live.

We do not need ease and safety; we need challenges commensurate with our abilities. We need, in Sir Walter Scott's words, a foeman worthy of our steel. We need a frontier. This is the world beyond door number two.

## Paradigm Shift

It is traditional, since Clarke's *Profiles of the Future*, to begin books of futurism like this one with lists of the short-sighted pronouncements of the "They said it couldn't be done" type, and then to list the cases where it had, in fact, been done. The one invariant in futurism before roughly 1970 was that predictions of social change overestimated, and of technological change underestimated, what actually happened. This invariant has been broken. With the notable exception of information technology, technological change has slowed and social change has mounted its crazy horse and galloped madly off in all directions. H. G. Wells's image of the feckless Eloi, a lampoon of the effete idle English upper classes of a century ago, describes many of us better than he could have imagined. We have dropped the ball (or had it knocked from our grasp) on most of our opportunities over the past half-century. But it remains perfectly possible, technologically, to regain the pace and have a *Jetsons* world by 2062—if we want to.

The main phases of technology that drove the Industrial Revolution were, first, low-pressure steam engines, where the force was atmospheric pressure acting against a vacuum produced by condensing steam. These were good enough to power pumps, mills, looms, lathes, and other machinery. This led to machine tools, and then high-pressure steam engines enabled by the precision that the machine tools made possible. High-pressure steam, where the steam pushes rather than pulls, had the power-to-weight ratios that allowed for engines in vehicles, notably locomotives and steamships. The energy in coal was critical for its power-to-weight ratio, and incidentally prevented the complete deforestation of England. Powered factories and commerce led to something like a two-orders-of-magnitude improvement in what the average person could have or do, and the greatest improvement in the general quality of life that the human race has ever seen.

To get an intuitive grasp of what the Industrial Revolution has meant to our lives, let us run it backward and see what we lose.

First of all, 90 percent of us would starve. World population in 1700 was only a tenth of today. Without industrial-scale production of ammonia-based fertilizer, the arable land on the surface of the Earth is nowhere near productive enough to feed us. Without modern freight transport, local famines would (and once did) happen at the whim of the weather.

The people left would only live half as long, on average, as we do; illness and injury were often deadly. I wouldn't be here: I was saved from acute bronchitis by an oxygen tent at the age of two.

On the upside, the average man was two or three times more physically strong than his modern-day counterpart. But that was because virtually everything that was done was done by muscles, mostly human ones. Almost everyone was a dirt farmer, plowing behind a mule all day in the baking sun in a cloud of flies, or cutting wood in the freezing cold. Without underwear. Or she was the dirt farmer's wife, who worked long hours drawing and carrying water, splitting wood, killing and cleaning chickens and rabbits, cooking, cleaning, making clothing, and washing it sparingly with skin-burning homemade lye soap. Not to mention giving birth without anything even vaguely resembling what we would call decent medical attention, and thus dying, on the average, younger than the men.

Candles are likewise expensive; your privy is outdoors (and unheated); you probably don't own a book or have all your teeth; you haven't traveled more than a few miles from where you were born, and when you did travel you walked. The 1 percent in your world don't have a private jet; they have a coach and four.

Compare that with your present life: a car; restaurants five minutes away; indoor plumbing, heating, and cooling; electric lights that turn on with the flip of a switch; a stove and microwave that do the same; a closet full of varied, comfortable clothing; a washer and dryer; a Roomba; a computer; a smart TV on which you can instantly access literally millions of videos, video chat with friends all over the world, and pull up literally millions of books in a minute; and the ability to order any of literally millions of products and have them arrive within a few days.

Now run the same difference *forward*. You may have trouble envisioning a fully fleshed-out world after another Industrial Revolution. I certainly do, and I've been trying for some time.

The three major, interacting, and mutually accelerating technologies in the 21st century are likely to be nuclear, nanotech (biotech is the "low-pressure steam" of nanotech), and AI, coming together in a synergy that I have taken to calling the Second Atomic Age. But you might as well call it the Second Industrial Revolution. It could easily improve productivity another couple orders of magnitude. It could almost certainly lengthen our Lives beyond the antediluvian Standard. The science fiction writers of the 1950s imagined mastering the solar system with a slightly advanced version of mature Industrial Revolution technology, but Second Atomic Age technology is much more likely the level needed to make this mastery feasible, viable, and sustainable.

## Technium

Explanations for the Great Stagnation come in two flavors. We can compare them using the concept of the "technium," Kevin Kelly's term for the sum total of the things that we can make and do, growing and interdependent like an ecosystem. We can imagine the technium to be like a water level, and the landscape about us to be determined by physics and economics. We live on the shores and sail on the waters of the technium. At first, simple techniques cause the habitable landscape to be cramped. The shores of Threadspin Pool give us wool tunics after three months of labor; but when the waters of the technium have risen to fill the channel of the River Loom and its tributaries, the valley of the Loom is wide, rich, and fertile.

The default Stagnation explanation has it that the technium has been rising since the 1970s, but the landscape has been a barren high desert, with no lowlands or valleys. The only exception was the Computer Sea, and all of the rising waters have poured into that. On this view, which is, in some sense, the same story as the low-hanging fruit, there are simply no other valleys out there. I began this investigation believing that might be the explanation, save for a few exceptions I knew about. But I have concluded that it is wrong. There are valleys galore, waiting to be made fertile by the waters of the technium.

In the Biblical account, the Garden of Eden was a wonderful fertile valley, but our ancestors were driven out of it by an angel with a flaming sword, simply because they desired knowledge. I have come to believe that the cause of stagnation is not a lack of valleys but an overabundance of angels with flaming, passionate, religious fervor for expelling people from paradise, lest they obtain forbidden knowledge.

In the Genesis story, the angel remained at the entrance of Eden to prevent the return of Adam and Eve. But, in reality, there are many ways to get into valleys. The rising waters of the technium can flow in from any connecting channel or low-lying pass. On that view, there are more and more viable passes into the valley as the technological level rises. Sometimes that takes the form of a technological end run, as in the current drone revolution. Sometimes it simply involves someone escaping the ban. Or, if the sclerotic polity is but one of many, the rest of the world will simply leave it behind, as happened to shogun-era Japan. Sometimes the political winds shift, as has happened in modern-day China. But, in every case, the farther the technium rises, the less secure the empty valley becomes.

Over the long course of history, technology has led science. Tinkerers find things that work, and scientists come along and explain them, in the process laying the groundwork for more thorough exploitation of the principles involved. People had been burning wicks in oil and tallow wax for millennia before Faraday's famous lectures on the candle flame. The makers of wine and cheese had, likewise, been using microbes long before Pasteur identified them. Steam engines had been in use for a century before Carnot elucidated the principles of thermodynamics.

Yes, we should have flying cars. Yes, we should have power too cheap to meter. Yes, we should have orbital hotels and a base on the Moon. Average family income in the U.S. should be $200,000 by now, and growing at a sustained 6 percent. But what has happened is that cultural reaction and regulatory ossification have combined to dam up the normal flow of experimentation in high-power technology. Where the technium would have spilled into the fertile valleys, we have instead built up a theoretical, scientific overhang.

We built and tested molten salt reactors and nuclear rocket engines over 40 years ago, but in the interim the United States has effectively used experimentation with nuclear capability as a *casus belli* against anyone who has tried it. We could have been developing nanotech since 1960, when Feynman first pointed out the possibilities. We know a lot more about the molecular scale now, so when the dam breaks it will come a lot faster. We have system designs for self-replicating nanofactories. We have detailed, atom-for-atom designs and engineering analyses of gears, levers, bearings, shafts, and sprockets. We can already simulate the quantum electrodynamics on our almost ridiculously powerful computers. It is as if we had gone into the Industrial Revolution already knowing thermodynamics and high-temperature metallurgy.

Should the cultural pendulum swing back, should we lose our idiotic fear of energy and regain the Henry Adams Curve, or should there arise elsewhere in the world a culture that has the same innovative spirit that we had just a century ago, huge technological advances could happen almost overnight as experimentation regains the lead, using all the "pent-up knowledge" of the past half-century as the overhang collapses, the dam breaks, and the technium surges into the forbidden valleys.

Nuclear power and nanotech are the most obvious examples of overhang technology; many others can be found by filling in the top right half of the expected-vs.-achieved technologies chart. (See Figure 8 / Appendix A.) You will recall that the technologies not achieved had a strong bias toward high energy intensity. Space travel is a prominent thread among them. High-power technologies promote an active frontier, be it the oceans or outer space. Frontiers, in turn, suppress self-deception and virtue-signaling in the major institutions of society, with its resultant cost disease. We have been caught, to some extent, in a self-reinforcing trap, as the lack of frontiers foster those pathologies, which limit what our society can do, including exploring frontiers. But, by the same token, we should also get positive feedback by going in the opposite direction, opening new frontiers and pitting our efforts against nature instead of each other.

Breakthroughs in the overhang technologies would allow the rest of the world to catch up with stagnating Western civilization, in a rerun of the collapse of Pax Britannica.

Weapons also have a strong high-power bias. Nanotech would enable cheap home isotopic separation. Short of that, it would enable the productivity of the entire U.S. military-industrial complex in an area the size of, say, Singapore. It's available to anyone who has the sense to follow Feynman's pathway and work in productive machinery instead of ivory-tower tiddlywinks. The amount of capital needed for a decent start is probably similar to that required by a well-equipped dentist's office. Cold fusion, HB11 fusion/fission, or any other nuclear power breakthrough that enables vehicle-sized reactors or nuclear-density batteries would have a similar effect.

The science fiction of the 1940s and 1950s was written in a world of burgeoning aerospace capabilities which were reflected in its visions of the future. Similarly, they assumed the future would be powered by nuclear energy, or more precisely by engines that would stand in the same relation to the "atomic piles" of the day as gas turbines do to medieval black-powder cannon.

The future isn't what it used to be. The hot technologies of today are computing, communication, and biotechnology. Modern science fiction (and futurism) has a tendency to predict a future along just as simple a linear extrapolation as did *The Jetsons* in its day. But it would be a major Failure of Nerve, much less of Imagination, to assume that the course of technology has ceased to change.

Most of the technologies that I've examined here are those we could have had by now, had we not dropped the ball. Second Atomic Age technology is not something beyond our grasp. It is something that we have been too busy "being Wolves to each other" to get to work and build. But let us imagine that, by some miracle, we manage to find Benjamin Franklin's "moral Science," perhaps something that comes out of our efforts to build consciences for robots, and that we get back on the Henry Adams Curve and regain our legacy as a productive, growing, improving, and optimistic culture.

As a technologist, I have in this book tended to concentrate on physical invention. But value, the gap between how well off we are and how much better we might be, is the pull to invention's push. The quality of governance is probably the strongest case in point. Technologically, the whole world could be at Rosling's Level 4 right now, but instead it is spread out in a slightly down-skewed distribution from 4 to 1. At a weighted average, the world could be 16 times as wealthy as it is if only our political systems were honest and competent. The difference, 15 times the world's current output, is an enormous value, simply sitting there waiting to be reaped. When someone invents a method of turning a Nigeria into a Norway, extracting only a 1 percent profit from the improvement, they will become rich beyond the dreams of avarice, and the world will become a much better and happier place.

Turning a Level 4 world into a Level 5 world (bringing along with the change flying cars) could be worth very roughly another 50 times the current total output of the human race. That should be a fair incentive. And remember, the original Industrial Revolution didn't stop after just one level.

## Futures Past

Philip K. Dick, an American science fiction writer of the mid-20th century, wrote the stories that became the movies *Blade Runner*, *Minority Report*, and *Total Recall*, among others. Dick's stories were bereft of the optimism of golden-age science fiction; they lacked a sense of wonder; there were no inspiring bright ideas; they were dystopian. Dick himself was an amphetamine addict, attempted suicide, and ultimately died at the age of 53. And yet his science fiction has had an enormous influence, especially among the literati. Jill Lepore, Harvard historian, writes of the trend,

> Dystopia used to be a fiction of resistance; it's become a fiction of submission, the fiction of an untrusting, lonely, and sullen twenty-first century, the fiction

of fake news and infowars, the fiction of helplessness and hopelessness. It cannot imagine a better future, and it doesn't ask anyone to bother to make one. It nurses grievances and indulges resentments; it doesn't call for courage; it finds that cowardice suffices. Its only admonition is: Despair more.[264]

We can describe the Great Stagnation with a simple short quip: We are not living in the future of Robert Heinlein; we are living in the future of Philip K. Dick.

Pessimism has always been intellectually fashionable. A century before Dick drew breath, John Stuart Mill observed that "not the man who hopes when others despair, but the man who despairs when others hope, is admired by a large class of persons as a sage."[265] Thomas Macaulay, the great historian of England, had written in 1830:

> On what principle is it that, when we see nothing but betterment behind us, we are to expect nothing but deterioration before us? If we were to prophesy that in the year 1930 a population of fifty million, better fed, clad, and lodged than the English of our time, will cover these islands, that Sussex and Huntingdonshire will be wealthier than the wealthiest parts of the West Riding of Yorkshire now are, that machines constructed on principles yet undiscovered will be in every house, many people would think us insane.[266]

And, of course, Macaulay was exactly right, and the doomsayers spectacularly wrong. This was the Victorian era, the Industrial Revolution, the greatest sustained improvement of human conditions in history.

Science fiction has a long and valuable history of providing us with visions of a better world. Verne, Wells, Burroughs, Hugo Gernsback—even Bellamy. And Campbell, Doc Smith, van Vogt, Heinlein, Asimov, Randall Garrett, Piper, Niven, and Jerry Pournelle provided people with places and lives that they could imagine and aspire to create. Science fiction since the 1960s has signally failed in that regard: We have been fed, by and large, a diet of Chicken Little soup, ladled out over leg of Frankenstein. By this Faustian bargain, science fiction gained academic respectability and climbed out of its literary ghetto. It became a literature not of technological and scientific ideas, of adventure and hope, but of ivory-tower snobbery, looking down on ordinary people and their aspirations for flying cars and a better future. Decadent ergophobia is ever so much more stylish and avant-garde. Can science fiction regain its soul, and we the future we were promised?

Of course, in what might be considered the bellwether work of science fiction, *The Time Machine*, H. G. Wells predicts the slow, inevitable decline of the human race into the Eloi. Most writers ignored that as having little relevance to the near-term future. But Robert Heinlein, for one, took it as a challenge. Science fiction is, after all, a literature of ideas.

Heinlein's most difficult, and most misunderstood, novel, *Beyond This Horizon*, was an attempt to refute Wells. He accepted Wells's premise: "What, unless biologi-

cal science is a mass of errors, is the cause of human intelligence and vigour?"
*Horizon* opens:

> Their problems were solved: the poor they no longer had with them; the sick,
> the lame, the halt, and the blind were historic memories; the ancient causes of
> war no longer obtained; they had more freedom than Man has ever enjoyed.
> All of them should have been happy.[267]

Heinlein brings two, or perhaps three, possibilities to combat the inevitable decline. The first is that we arrange society to reduce safety and enhance the advantage
of intelligence and vigor in the evolutionary struggle. In the novel, the culture expects
young men to duel. It is a sad commentary on our culture today that the vast majority
of reviewers cannot even understand why it's being done, but even more so that they
don't realize that, in such a culture, the intelligence to avoid duels is at least as important as quick reflexes once you get into one.

Heinlein's second possibility was the most innovative. Writing a decade before
the discovery of DNA, he posited that we might be able to use technology to select the
genes in our offspring. Not only did he see the vast dangers of such a capability, but he
found a way past it, an ingenious technique that to this day has come to be called "the
Heinlein Solution." (In simple terms, when a couple goes to a genetic specialist to
have an improved baby, they are allowed to select among the genes they bear but not
introduce new ones, so their baby is one that they could have possibly had naturally.)
Heinlein is likely to cast as long a shadow over human genetic ethics as Asimov does
over robotics.

Heinlein's third possibility is narratively the weakest, involving the horizon of the
title, beyond which lie astral planes of telepathy, reincarnation, and other pseudo-reli-
gious phenomena. But it could easily be reduced to this: There are more important
things for the human race to be doing than eating and sleeping in comfort. And this is
inescapably and obviously true. We need to understand how to harness our enormous
creative energies for improvement instead of virtue-signaling. We need to understand
how to live beyond Earth's fragile soap bubble of a biosphere. We need to begin to appreciate how much more there is to even our small solar system.

We need to disturb the universe.

## Looking Upward

We have allowed Eloi angst to drive us into a moral funk about the true value of
humanity and science. But this is the kind of thing that can change.

Two quotations from Henry Adams will illustrate. First, from 1862:

> Man has mounted science, and is now run away with. I firmly believe that be-
> fore many centuries more, science will be the master of men. The engines he
> will have invented will be beyond his strength to control. Someday science

may have the existence of mankind in its power, and the human race commit suicide, by blowing up the world.[268]

And then, from 1907:

> He could see that the new American—the child of incalculable coal-power, chemical power, electric power, and radiating energy, as well as of new forces yet undetermined—must be a sort of God compared with any former creation of nature. At the rate of progress since 1800, every American who lived into the year 2000 would know how to control unlimited power. He would think in complexities unimaginable to an earlier mind. He would deal with problems altogether beyond the range of earlier society.[269]

These reflect drastically different takes on the possibilities of technology, filtered through the zeitgeist of different times. But in Adams's case they had shifted from a pessimistic view during the Civil War to an optimistic one in the halcyon days of the Edwardian decade. Twilight falls, but daybreak follows.

The Great Stagnation was the Qing Dynasty self-strangulation, rerun at internet speed.

The internet itself is the printing press of the Information Age; remember that the printing press was responsible not only for the rise of science but for centuries of religious wars as people escaped the yoke of the medieval church's centralized information control. Remember that the medieval papacy was venal, grasping, and entrenched even while it was held as sacred and infallible by a large segment of the people. Our monstrous bureaucracy today is no different. And yet the Reformation happened. It is history that teaches us to hope.

In many ways, the Second Atomic Age could far exceed what science fiction writers, much less "respectable" futurists, predicted. They knew that we could build crystal cities high above the scorched plains of Texas; few of them realized that we might convert those plains into a Big Sur or Yosemite at will. They knew that we could have flying cars; few realized we might remake ourselves so that we might fly unassisted. They knew that we could walk on other worlds; few realized that we could build other worlds, or live at ease and in comfort in space itself.

This will not have happened by 2062, but we could be well on our way already. Again, your model is 50 years of progress in physical machinery to match the 50 years we just had in computers and electronics. You have not only a self-driving flying car but a full domestic staff of robots. You can obtain most gadgets, clothing, and food by downloading designs, as you would a software app or movie today, and printing them on a synthesizer. Your life expectancy is well over 100, and rising at more than a year per year. Top athletes (and research mathematicians) retire at 80 instead of 40. No machine that you make or use needs fuel, nor burns any; but, of course, people still like fireplaces, not to mention getting together socially to eat and drink. There are many more cities on, and under, the sea than on the Moon, but there are a few of the latter as well.

And, of course, for any of this to have happened, your culture must view regulation the same way that ours views chattel slavery, and scientific bureaucracy the way that ours views eugenics.

Even this vastly understates the possibilities of a century on. It avoids Failure of Nerve, but we must at least try to confront Failure of the Imagination as well. Straight-forward projections of what we know technology should be capable of will pale in the face of fundamentally new science. If in 1900 you had predicted the technology of to-day, you would have had to do it without any knowledge of relativity or quantum me-chanics; of polymer chemistry or DNA; of electron shells, the atomic nucleus, neu-trons, or fission or fusion; of the basic theorems of computational science or information theory; of the basic structures of self-reproducing machines that match those in the biological cell; of transistors or, indeed, even vacuum tubes, to say noth-ing of microwaves and radar.

Do you think that science has *stopped?* Seriously?

Tom Swift Jr. had a Flying Lab—a huge airplane able to go anywhere in the world, packed with every kind of equipment that a boy genius could want, from chemistry glassware to machine tools. It even carried a small helicopter to buzz around in after you landed somewhere. How many things could you discover or invent if you had something like this?

You do have something like this now: You can "go" all over the world on the inter-net, talk to any of a vast number of people if you wish. You have access to more comput-ing power and design and simulation software than existed in the entire world in 1962. You can experiment, you can tinker, you can invent, you can discover, you can learn.

The Second Atomic Age story shows us that this can happen in the world of atoms as well as the one of bits. A real flying lab, if you want, but at any rate a lab, a machine shop, a 3D printer capable of making more things than you can imagine, and in par-ticular capable of making something to improve the world that only you have imag-ined. Your lab is not just full of machines but staffed by intelligent robots that are ex-pert in the full history of science and engineering.

If we did get back on the Henry Adams Curve, with energy and population growth in their historic proportions, by 2200 or so we would reach the status of a Kardashev Type I civilization, with 100 times today's population and 100 times today's power per capita. (That's only a 2.5 percent growth rate.) This would be a disaster if Earthlings all lived on the Earth's surface, but would be just about right for a broadly based solar-system-scale civilization. The resumption of a dynamic ethic in an open frontier could serve to reduce the primacy of virtue signaling and cost disease in our major institutions.

The key to that future will be our visions: whether we can imagine, want, and try hard enough to achieve great things—things worthy of the capabilities of a people with intelligent robots, complete control of the structure of matter, and the limitless power of the atom. Our muscle awaits, if only we can find our muse. We need hopers and dreamers; we need visionaries who can see a better future worth striving for. We need great, important things to do with the staggeringly huge capabilities that lie within our grasp. Science fiction must get back down into the gutter and start looking back up at the stars.

> *It is not really necessary to look too far into the future; we see enough already to be certain it will be magnificent. Only let us hurry and open the roads.*

> —**Wilbur Wright**

# Energy Intensity of Predicted Technologies

These are the technologies expected or predicted by the science fiction writers or futurists in the '60s, from the list given in Chapter 1.

If a technology continued developing over the whole 50 years, I rated it by an estimate of how well it is doing now compared to what the writers expected. On the other hand, if a technology flatlined, we can look at what percentage of the past 50 years it kept developing on an expected schedule.

Energy intensity is log base 10 of watts: A human at subsistence is 100 watts, or 2; an American today (or in 1975!) is 10 kilowatts, or 4. A car gets rated 4 even though it may have a multi-hundred-kilowatt engine, because at average cruise it is only putting out on the order of tens. An airplane with the same nominal horsepower gets a 5, though, because you run the airplane engine at a higher power level most of the time.

Admittedly, the coordinates for each technology are not only estimates but somewhat subjective estimates. Even so, I was quite surprised at the shape of the resulting graph; I had expected only a mild correlation, a blob slightly tilted to the left. (See Figure 51.) I have refrained from updating any estimate since seeing the graph.

- **Pocket telephones: 200 percent; 1.** Not only are they more common now than most SF writers would have predicted, but they do more. They use little energy, even prorating your share of the tower and network.

- **Home-based videophones (also used for reading and commerce): 100 percent; 2.** Note that this list was made before the pandemic and Zoom revolution; I might give it a stronger achieved rating now.

- **Space stations: 40 percent; 6.5.** We have one, but they expected more, with commercial traffic. Note that there is a decent chance that this one will pick up again in the coming decade.

- **Lunar landings: 15 percent; 7.** Not only did we flatline, we zeroed.

- **Nuclear rockets: 5 percent; 9.**

- **Lunar base: 0 percent; 8.**

- **Interplanetary travel: 0 percent; 9.** The writers were thinking about manned landings. Note that Asimov expected a lot of robots to be involved in the exploration of the solar system, but, in reality, sending robots is an order of magnitude easier than sending humans whom you need to get back home!

- **Interplanetary colonies: 0 percent; 10.** This is virtually impossible without nuclear rockets.

- **Cars: 100 percent; 4.**

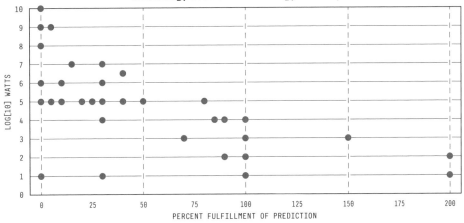

Figure 51: Kardashev's scale. Power numbers are log watts.

- **Self-driving cars: 100 percent; 4.** We got just about as far toward them as the best-informed guesses.

- **Hovercraft: 10 percent; 5.** A few sport vehicles and ferries, but no major personal transportation.

- **Private planes: 30 percent; 5.** (i.e., up to the collapse of the industry.)

- **Convertible planes: 10 percent; 5.**

- **VTOL flying cars: 0 percent; 6.**

- **Transportation 1,000 miles per hour and one cent per mile: 30 percent; 6.** We got half the speed, and a mile costs a nickel.

- **Highways in decline as air vehicles predominate: 20 percent; 5.5.** (as an average for air travel.) Highways are in decline, but mostly from mismanagement rather than air travel.

- **Batteries: 70 percent; 3.** Chemical batteries have come a long way; the writers expected electric cars.

- **"Atomic" batteries: 0 percent; 5.** The reasonable expectation was for something 100 to 1,000 times as good as we have now.

- **Fusion power: 0 percent; 6.** "Too cheap to meter."

- **Wireless energy transmission: 10 percent; 5.**

- **Space-based military: 30 percent; 7.** Nuclear ICBMs are the defining background of all things military, and space-based recon is common. But the writers expected military space stations and Moon bases.

- **Cyborgs: 50 percent; 4.** Mostly for medical instead of enhancement.

- **Robots: 100 percent; 3.** Asimov nailed the current status with "will neither be common nor very good in 2014, but they will be in existence."

- **Automation eliminating jobs: 50 percent; 4.**

- **Household robots: 90 percent; 2.**

- **Industrial robots: 100 percent; 3.** The difference between this and "automation" has to do with the expectations associated with the terms.

- **Translating machines: 90 percent; 4.** Workable translation is a "big data" phenomenon and implies a data center, not a PC.

- **Artificial intelligence: 80 percent; 5.** Even bigger data center; they expected something a solid notch more capable than Alexa.

- **Global library: 150 percent; 3.** They expected a library to be shared, unlike an AI.

- **Computers: 200 percent; 2.** This one is your PC or tablet.

- **"Atomic" power: 25 percent; 5.** (with the flatline.)

- **Undersea cities: 0 percent; 6.**

- **Floating cities: 20 percent; 5.** Think cruise ships.

- **Domed cities: 5 percent; 5.** Mostly stadiums.

- **Sea mining: 10 percent; 6.** Mostly oil wells.

- **Innovative housing technology: 40 percent; 5.**

- **Automatic home meal preparation: 30 percent; 3.** Mostly the microwave and beverage makers combined with prepackaged meals.

- **Major substitution of synthetic food for agriculture: 50 percent; 5.** I'm taking this category to include the industrialization and energy intensification of agriculture itself, which is clearly moving in the direction of factory-produced food.

- **Contraception, revising relations between the sexes: 100 percent; 1.**

- **Cures for cancer, the common cold, and tooth decay: 30 percent; 1.**

- **Psychology and education a hard science: 0 percent; 1.**

# Selected Readings

Sources for background and further reading. I have tried to include much of what I consider "classic" science fiction, i.e., which would have contributed to the overall sense of "the future we were promised." In many cases, with nonfiction I have included books that are readable rather than more voluminous academic tomes (the equivalent of, e.g., Toynbee's *Civilization on Trial* rather than his *Study of History*). Except for the FAR/AIM, you might reasonably expect to read or to have read all of these. This does not include scientific papers, which are referenced in footnotes. If you want to go deeper into any of these subjects, the books themselves, not to mention Google, will be your guide.

### Chapter 1: The World of Tomorrow
- Victor Appleton II, *Tom Swift and His Triphibian Atomicar* (Grosset and Dunlap, 1962).
- Hugo Gernsback, *Ralph 124C 41+* (first serialized 1911), https://www.gutenberg.org/ebooks/60944.
- Arthur C. Clarke, *Profiles of the Future* (Pan, 1962).
- H. G. Wells, *Things to Come*, film, (United Artists, 1936).
- Isaac Asimov, *Nine Tomorrows*, (Doubleday, 1959).
- Robert A. Heinlein, *Podkayne of Mars* (first serialized 1962; Putnam, 1963).
- Herman Kahn and Anthony J. Wiener, *The Year 2000: A Framework for Speculation on the Next Thirty-Three Years* (MacMillan, 1967).
- Gregory Benford, *The Wonderful Future That Never Was: Flying Cars, Mail Delivery by Parachute, and Other Predictions from the Past* (Hearst, 2012).
- David Gelernter, *1939: The Lost World of the Fair* (Harper, 1995).

### Chapter 2: The Graveyard of Dreams
- Tyler Cowen, *The Great Stagnation* (Dutton, 2011).
- Robert Prehoda, *Designing the Future* (Chilton, 1967).
- Robert Gordon, *The Rise and Fall of American Growth* (Princeton University, 2016).
- Philip Francis Nowlan, *Armageddon 2419 A.D.* (1928), https://gutenberg.org/ebooks/32530.
- Robert Cartmill, *The Next Hundred Years . . . Then and Now* (Xlibris, 2002).
- Geoffrey Hoyle, *2011: Living in the Future* (first published 1972; Laughing Elephant, 2010).
- Mancur Olson, *The Rise and Decline of Nations* (Yale University, 1982).
- Henry Adams, *The Education of Henry Adams* (1907), https://www.gutenberg.org/files/2044/2044-h/2044-h.htm.

### Chapter 3: The Conquest of the Air
- Jules Verne, *Robur the Conqueror* (1886), https://www.gutenberg.org/ebooks/3808.
- Juan de la Cierva, *Wings of Tomorrow: The Story of the Autogiro* (Brewer Warren & Putnam, 1931).

- Frank Kingston Smith, *Legacy of Wings: The Harold F. Pitcairn Story* (Rowman & Littlefield, 1981).
- Patrick Gyger, *Flying Cars: The Extraordinary History of Cars Designed for Tomorrow's World* (Haynes, 2011).
- Andrew Glass, *Flying Cars: The True Story* (Clarion, 2015).
- George Green, *Flying Cars, Amphibious Vehicles and Other Dual Mode Transports: An Illustrated Worldwide History* (McFarland & Company, 2010).

## Chapter 4: Waldo and Magic, Inc.
- Robert Heinlein, *Waldo & Magic, Inc.* (Doubleday, 1950). (The story "Waldo" was originally published in *Astounding Magazine*, 1942).
- K. Eric Drexler, *Engines of Creation: The Coming Era of Nanotechnology* (Anchor, 1986); *Nanosystems: Molecular Machinery, Manufacturing, and Computation* (Wiley, 1992).
- J. Storrs Hall, *Nanofuture: What's Next for Nanotechnology* (Prometheus, 2007).

## Chapter 5: Cold Fusion?
- E. E. "Doc" Smith, *The Skylark of Space* (1928), https://www.gutenberg.org/ebooks/20869.
- John Huizenga, *Cold Fusion: The Scientific Fiasco of the Century* (University of Rochester, 1992).
- Charles Beaudette, *Excess Heat: Why Cold Fusion Research Prevailed* (Oak Grove, 2000).

## Chapter 6: The Machiavelli Effect
- Edgar Rice Burroughs, *The Gods of Mars* (1913), https://www.gutenberg.org/ebooks/64.
- David Freedman, *Wrong: Why Experts Keep Failing Us—and How to Know When Not to Trust Them* (Little, Brown & Company, 2010).
- Kevin Kelly, *What Technology Wants* (Penguin, 2010).
- Bryan Caplan, *The Case Against Education: Why the Education System Is a Waste of Time and Money* (Princeton University, 2018).
- Steven Johnson, *Where Good Ideas Come From: The Natural History of Innovation* (Riverhead, 2011).
- Matt Ridley, *How Innovation Works: And Why It Flourishes in Freedom* (Harper, 2020).
- Terrence Kealy, *The Economic Laws of Scientific Research* (Palgrave MacMillan, 1996).

## Chapter 7: The Age of Aquarius
- H. G. Wells, *The Time Machine* (1895), https://www.gutenberg.org/ebooks/35.
- Robert Trivers, *Deceit and Self-Deception: Fooling Yourself the Better to Fool Others* (Allen Lane, 2011).
- Kevin Simler and Robin Hanson, *The Elephant in the Brain: Hidden Motives in Everyday Life* (Oxford University, 2017).
- Donella H. Meadows, Dennis L. Meadows, Jørgen Randers, William W. Behrens III (commissioned by the Club of Rome), *Limits to Growth* (Potomac, 1972).
- Jerry Pournelle, *A Step Farther Out* (Ace, 1979).
- Julian Simon, *The Ultimate Resource* (Princeton University, 1981).
- Matt Ridley, *The Origins of Virtue: Human Instincts and the Evolution of Cooperation* (Viking, 1996).

## Chapter 8: Forbidden Fruit
- Alex Tabarrok, *Launching the Innovation Renaissance: A New Path to Bring Smart Ideas to Market Fast* (TED Books, 2011).
- Austin B. Hallock, *Sky Full of Dreams: The Aviation Exploits, Creations, and Visions of Bruce K. Hallock* (Elevon Books, 2010).

- Philip K. Howard, *The Death of Common Sense: How Law Is Suffocating America* (Random House, 1994).
- Arthur C. Clarke, *Childhood's End* (Pan, 1956).
- H. G. Wells, *Anticipations of the Reaction of Mechanical and Scientific Progress Upon Human Life and Thought* (1901), https://www.gutenberg.org/files/19229/19229-h/19229-h.htm.
- D. S. L. Cardwell, *Turning Points in Western Technology* (Heinemann, 1972).
- Hans Rosling, *Factfulness: Ten Reasons We're Wrong About the World—and Why Things Are Better Than You Think* (Flatiron, 2018).

**Chapter 9: Ceiling and Visibility Unlimited**
- Orville and Wilbur Wright, *The Early History of the Airplane* (1922), https://www.gutenberg.org/ebooks/25420.
- John Denker, *See How It Flies* (1996), https://www.av8n.com/how/.
- H. G. Wells, *The War in the Air* (1908), https://www.gutenberg.org/ebooks/780.
- James Tobin, *To Conquer the Air* (Free Press, 2004).
- Antoine de Saint-Exupéry, *Wind, Sand, and Stars* (Reynal and Hitchcock, 1939).
- Ernest K. Gann, *Fate Is the Hunter* (Stratford, 1961).
- Richard Bach, *Jonathan Livingston Seagull* (MacMillan, 1970).

**Chapter 10: Dialogue Concerning the Two Great Systems of the World**
- Steve Markman and Bill Holder, *Straight Up: A History of Vertical Flight* (Schiffer, 2000).
- James R. Chiles, *The God Machine: From Boomerangs to Black Hawks: The Story of the Helicopter* (Random House, 2007).
- Bruce Charnov, *From Autogiro to Gyroplane: The Amazing Survival of an Aviation Technology* (Praeger, 2003).

**Chapter 11: The Atomic Age**
- H. G. Wells, *The World Set Free* (1914), https://www.gutenberg.org/ebooks/1059.
- Richard Rhodes, *The Making of the Atomic Bomb* (Simon & Schuster, 1986).
- Bernard L. Cohen, *The Nuclear Energy Option: An Alternative for the 90s* (Springer, 1990).
- L. Sprague de Camp, *Man and Power: The Story of Power From the Pyramids to the Atomic Age* (Golden, 1961).

**Chapter 12: When Worlds Collide**
- Philip Wylie and Edwin Balmer, *When Worlds Collide* (Frederick A. Stokes, 1933).
- Paul Alivisatos et al., *Productive Nanosystems: A Technology Roadmap,* (Battelle Memorial Institute and Foresight Nanotech Institute, 2007).
- Robert Freitas and Ralph Merkle, *Kinematic Self-Replicating Machines* (Landes Bioscience, 2004), http://www.MolecularAssembler.com/KSRM.htm.
- R. A. Freitas and W. P. Gilbreath, *Advanced Automation for Space Missions* (NASA CP 2255, 1980).
- Neil Gershenfeld, *Fab: The Coming Revolution on Your Desktop—From Personal Computers to Personal Fabrication* (Basic, 2005).

**Chapter 14: The Dawn of Robots**
- Karel Čapek, *R.U.R.* (play, 1920), https://www.gutenberg.org/files/59112/59112-h/59112-h.htm.
- Isaac Asimov, *I, Robot* (Gnome, 1950).
- Erik Brynjolfsson and Andrew McAfee, *Race Against the Machine* (Digital Frontier, 2011).
- Randall Garrett, *Unwise Child* (Doubleday, 1962).

- David A. Mindell, *Our Robots, Ourselves: Robotics and the Myths of Autonomy* (Viking, 2015).
- J. Storrs Hall, *Beyond AI* (Prometheus, 2005).
- William Tenn, *Of Men and Monsters* (Ballantine, 1968).

**Chapter 15: The Second Atomic Age**
- Richard Rhodes, *Dark Sun: The Making of the Hydrogen Bomb* (Simon & Schuster, 1995); *Nuclear Renewal: Common Sense About Energy* (Viking, 1993).
- Robert Freitas, Jr., *Nanomedicine* (Landes Bioscience, 1999).
- K. Eric Drexler, *Radical Abundance: How a Revolution in Nanotechnology Will Change Civilization* (PublicAffairs, 2013).
- Mats Lewan, *An Impossible Invention* (Mats Lewan, 2014).

**Chapter 16: Tom Swift and His Flying Car**
- Theodore von Kármán, *Aerodynamics: Selected Topics in the Light of Their Historical Development* (Cornell University, 1954).
- Ira H. Abbott and A. E. von Doenhoff, *Theory of Wing Sections* (Dover, 1959).
- Henk Tennekes, *The Simple Science of Flight* (MIT, 1996).
- Barnes W. McCormick Jr., *Aerodynamics of V/STOL Flight* (Academic, 1967).
- Daniel Raymer, *Simplified Aircraft Design for Homebuilders* (Design Dimension, 2002).

**Chapter 17: Escape Velocity**
- Arthur C. Clarke, *The Promise of Space* (Harper & Row, 1968).
- Tom Wolfe, *The Right Stuff* (Farrar, Straus & Giroux, 1979).
- Robert A. Heinlein, *Space Cadet* (Ace, 1948).
- Harold L. Goodwin (as Blake Savage), *Rip Foster Rides the Gray Planet* (Whitman, 1952).
- Donald A. Wollheim, *The Secret of Saturn's Rings* (John C. Winston, 1954).
- Eugene Cernan and Don Davis, *The Last Man on the Moon: Astronaut Eugene Cernan and America's Race in Space* (St. Martin's, 1999).
- Gerard O'Neill, *The High Frontier: Human Colonies in Space* (William Morrow & Co., 1976); *2081: A Hopeful View of the Human Future* (Simon & Schuster, 1981).
- George Dyson, *Project Orion: The True Story of the Atomic Spaceship* (Henry Holt, 2002).

**Chapter 18: Metropolis**
- Isaac Asimov, *The Caves of Steel* (Pyramid, 1953).
- Clifford Simak, *City* (Gnome, 1952).
- Robert Heinlein, *For Us, the Living* (Scribner, 2003). (The novella was originally written in 1938, but was not published until 2003.)
- Jane Jacobs, *Dark Age Ahead* (Random House, 2004).
- James A. Dunn Jr., *Driving Forces: The Automobile, Its Enemies, and the Politics of Mobility* (Brookings Institution, 1998).
- Randal O'Toole, *The Vanishing Automobile and Other Urban Myths: How Smart Growth Will Harm American Cities* (Thoreau Institute, 2000).

**Chapter 19: Engineers' Dreams**
- L. Sprague de Camp, *The Continent Makers* (1951); the novella is included in *The Continent Makers and Other Tales of the Viagens* (Twayne, 1953).
- Arthur C. Clarke, *The City and the Stars* (Muller, 1956).
- John W. Campbell, *The Mightiest Machine* (Hadley, 1947).
- Ben Bova, *The Weathermakers* (Winston, 1967).
- E. E. "Doc" Smith, *Spacehounds of IPC* (Fantasy, 1947).

**Chapter 20: Rocket to the Renaissance**

- Isaac Asimov, *Foundation* (Gnome, 1951).

- Jerry Pournelle, *A Step Farther Out* (W.H. Allen, 1980).

- Poul Anderson, *Brain Wave* (Ballantine, 1954).

- H. Beam Piper, *The Cosmic Computer* (Ace, 1963).

- Matt Ridley, *The Rational Optimist: How Prosperity Evolves* (Harper, 2010).

- Brink Lindsey, *Reviving Economic Growth: Policy Proposals From 51 Leading Experts* (Cato Institute, 2015).

- Ray Kurzweil, *The Singularity Is Near: When Humans Transcend Biology* (Viking, 2005).

# Epigraph Sources

168 Amory Lovins, "The Plowboy Interview with Amory Lovins," *Mother Earth News,* November/December 1977.

## Chapter 12: When Worlds Collide

170 Napoleon Bonaparte, a widely repeated quote from the time of the Napoleonic Wars, 1803–1815.

172 Arthur C. Clarke, *Profiles of the Future* (1962), Chapter 13: "Aladdin's Lamp."

175 K. Eric Drexler, Testimony before the Subcommittee on Science, Technology, and Space, U.S. Senate, June 26, 1992.

178 Seabees Motto, believed to have been coined during World War II. Variations have also been associated with the U.S. Army Corps of Engineers and the Army Air Forces during the same period.

181 H. G. Wells, *Anticipations of the Reaction of Mechanical and Scientific Progress Upon Human Life and Thought* (1902), footnotes to Chapter 3: "Developing Social Elements."

## Chapter 13: When the Sleeper Wakes

184 Sam Moskowitz, *The Immortal Storm: A History of Science Fiction Fandom* (1954), Chapter 1: "Introduction."

## Chapter 14: The Dawn of Robots

194 H. G. Wells, *Things to Come: A Film by H. G. Wells*, film script (Macmillan, 1935).

198 Robert Heinlein, *Time Enough for Love* (Putnam, 1973), "Intermission."

200 Philip Francis Nowlan, *Armageddon 2419 A.D.* (1928), Chapter 2: "The Forest Gangs."

207 Alan M. Turing, "Computing Machinery and Intelligence, *Mind* LIX, no. 236 (October 1950): 433–460.

## Chapter 15: The Second Atomic Age

211 Julian Simon, *The Ultimate Resource 2* (Princeton University Press, 1996), 181.

222 Henry Adams, *The Education of Henry Adams* (1907), Chapter 34: "A Law of Acceleration."

## Chapter 16: Tom Swift and His Flying Car

224 Arthur Sweetser and Gordon Lamont, *Opportunities in Aviation* (Harper & Brothers, 1920), Chapter 7: "Landing-Fields—The Immediate Need."

228 Ralph Waldo Emerson, possibly attributed posthumously.

## Chapter 17: Escape Velocity

238 Jules Verne, *From the Earth to the Moon: A Direct Route in 97 Hours, 20 Minutes* (1865), Chapter 19: "A Monster Meeting."

244 T. S. Eliot, "The Love Song of J. Alfred Prufrock" (1915).

247 Konstantin Tsiolkovsky, Personal correspondence, 1911.

## Chapter 18: Metropolis

250 Dennis Poon, quoted in Sophia Chen, "How to Keep a 1,500-Foot Skyscraper From Falling Over," *Wired*, July 27, 2015.

253 Jane Jacobs, *The Death and Life of Great American Cities* (Random House, 1961), "Introduction."

260 Syd Mead, in an interview with Patrick Sisson, *Curbed*, 2015.

## Chapter 19: Engineers' Dreams

264 Glenn Reynolds, remarks.

## Chapter 20: Rocket to the Renaissance

278 Benjamin Franklin, letter to Joseph Priestly, February 8, 1780.

290 Wilbur Wright, Gold Medal acceptance speech before the Aero Club of France, November 5, 1908.

# Endnotes

1   Victor Appleton II, *Tom Swift and His Triphibian Atomicar* (Grosset and Dunlap, 1962), 1-16.
2   E.g., Garrett P. Serviss, *Edison's Conquest of Mars*, serialized in the *New York Evening Journal*, 1898.
3   David Gelernter, *1939: The Lost World of the Fair* (Free Press, 1995), 343.
4   *Things to Come: A Film by H. G. Wells* (Macmillan, 1935). This is the script, by Wells, of the movie, derived from but not greatly similar to Wells's earlier novel *The Shape of Things to Come*.
5   Waldemar Kaempffert, "Green Light for the Age of Miracles," *The Saturday Review*, April 22, 1944, 14, http://www.unz.org/Pub/SaturdayRev-1944apr22-00014.
6   Arthur C. Clarke, *Profiles of the Future* (Pan, rev. ed. 1973), 48.
7   Isaac Asimov, "Visit to the World's Fair of 2014," *New York Times*, August 16, 1964, http://www.nytimes.com/books/97/03/23/lifetimes/asi-v-fair.html.
8   In a BBC *Horizon* interview in 1964, available at https://www.youtube.com/watch?v=KT_8-pjuctM.
9   "NERVA," Wikipedia, June 18, 2021, http://en.wikipedia.org/wiki/NERVA. The entry lists six successful tests from 1964 to 1969.
10  "Project Pluto," Wikipedia, May 12, 2021, http://en.wikipedia.org/wiki/Project_Pluto.
11  "New Frontier Mining Under the Sea," *Mining Technology*, November 25, 2008, http://www.mining-technology.com/features/feature46357.
12  The spelling was changed from "Rosey" in the 1980s reappearance of the series. She is clearly modeled on Hazel, the titular human maid of a contemporaneous live-action sit-com and earlier comic strip.
13  Notably, e.g., Robert Gordon, *The Rise and Fall of American Growth* (Princeton University Press, 2016).
14  Philip Francis Nowlan, *Armageddon 2419 A.D.*, 1928, https://www.gutenberg.org/files/32530/32530-h/32530-h.htm.
15  Mark Steyn, "Motown No More," *Orange County Register*, July 19, 2013.
16  Paul Graham, "The Hundred-Year Language," essay posted at http://www.paulgraham.com/hundred.html, adapted from the keynote talk at PyCon 2003. Also included in his book *Hackers and Painters*.
17  Robert Gordon, "Is U.S. Economic Growth Over?," NBER Working Paper 18315, August 2012, https://www.nber.org/system/files/working_papers/w18315/w18315.pdf.
18  Cowen has repeated this quip in every talk I've heard him give.
19  "Flying Car," Wikipedia, accessed 2020, https://en.wikipedia.org/wiki/Flying_car.
20  Adam Gurri, "The Internet Has Enriched Us More Than Flying Cars Would Have," The Ümlaut, October 27, 2014, https://theumlaut.com/the-internet-has-enriched-us-more-than-flying-cars-would-have-f90a90918b87.
21  Kevin Drum, "Why the Future Is Brighter Than You Think," *Mother Jones*, October 7, 2011, http://motherjones.com/kevin-drum/2011/10/future-of-innovation.
22  Peter Thiel, "The End of the Future," *National Review*, October 3, 2011.
23  "The Machine Stops" was first published in *The Oxford and Cambridge Review* in 1909, and between hard covers in 1928. It may be read online at http://www.visbox.com/pra-jlich/forster.html.
24  Arthur C. Clarke, *Profiles of the Future* (Chaucer Press revised edition, 1973), 98.
25  The full quote is "I am, and ever will be, a white-socks, pocket-protector, nerdy engi-neer—born under the second law of thermodynamics, steeped in the steam tables, in love with free-body diagrams, transformed by Laplace, and propelled by compressible

flow. As an engineer, I take a substantial amount of pride in the accomplishments of my profession," said in an address to the National Press Club, February 22, 2000.

26 A comprehensive overview can be found in *Slate Star Codex*, at http://slatestarcodex. com/2017/02/09/considerations-on-cost-disease.

27 Robert W. Prehoda, *Designing the Future: The Role of Technological Forecasting* (Chilton, 1967), 53. Emphasis on the entire quoted passage is in the original.

28 See the Federal Reserve Bank of St. Louis's chart, at https://fred.stlouisfed.org/series/ MEFAINUSA672N.

29 Henry Adams, *The Education of Henry Adams* (1907), https://www.gutenberg.org/ files/2044/2044-h/2044-h.htm. This was his autobiography, written somewhat unusually in the third person.

30 U.S. Energy Information Administration, "Energy Sources Have Changed Throughout the History of the United States," *Today in Energy*, July 3, 2013, http://www.eia.gov/todayinenergy/detail.cfm?id=11951.

31 Simon Newcomb, "Outlook for the Flying Machine," *New York Independent*, October 22, 1903, 2508, 2510-2511. (This was published just a few weeks before the Wright brothers' successful first flight!)

32 The civilian version, the Vought V-173, was flown by, among other test pilots, Charles Lindbergh.

33 National Advisory Committee for Aeronautics, *Report 431: Characteristics of Clark Y Airfoils of Small Aspect Ratios* (1932).

34 Bill Maloney, "Custer Channel Wing CCW-5 Specifications," Mid-Atlantic Air Museum, March 19, 2008, http://www.williammaloney.com/Aviation/MidAtlanticAirMuseum/ CusterChannelwing/index.htm.

35 Bruce Charnov, *From Autogiro to Gyroplane* (Westport, CT: Praeger, 2003), 43.

36 Amelia Earhart, "Your Next Garage May House an Autogiro," *Cosmopolitan*, August 1931. It is likely that the article was actually ghostwritten by Pitcairn's ad agency.

37 See Frank Kingston Smith, *Legacy of Wings* (a biography of Pitcairn), (T-D 1981), Chapter 19: Catastrophe at Croydon.

38 General Aviation Manufacturers Association, https://gama.aero/documents/statistical-data-1973/statistical_data_1973_498b99d81d.pdf, p. 7.

39 Buckminster Fuller, *Inventions: The Patented Works of R. Buckminster Fuller*, 1983.

40 "Cierva Patent No. 2,380,582.—The trial commissioner found: (1) Claims 1, 2, 3, 4, 5, 7, 8, 9, 12, 13, 16, 17, 18, 19, 20 and 21 infringed by the Vertol HUP-1 helicopter, the Vertol H-21B helicopter, and the McCulloch MC-4C helicopter, (2) Claims 6, 7, 8, 9, 12, 13, 16 and 17 infringed by the Bell HSL-1 helicopter, and (3) Claims 16 and 17 infringed by the Bell HTL-4 helicopter."—*Autogiro Co. of America v. United States*, 384 F.2d 391, 397–98 (Ct.Cl.1967).

41 Originally designed by our friend Charles Zimmerman.

42 https://www.si.edu/object/hiller-model-1031-1-flying-platform percent3Anasm_ A19610070000, also https://www.hiller.org/event/flying-platform.

43 From a placard on the surviving vehicle, as of 2017.

44 Robert Heinlein (writing as Anson MacDonald), "Waldo," *Astounding Magazine*, August 1942. The story is also included in *Waldo & Magic, Inc.* (Doubleday: 1950).

45 Richard P. Feynman, "There's Plenty of Room at the Bottom," first published in *Engineering and Science* XXIII, no. 5 (February 1960), https://resolver.caltech.edu/ CaltechES:23.5.1960. See also "'Plenty of Room' Revisited." *Nature Nanotech* 11, no. 781 (2009), https://doi.org/10.1038/nnano.2009.356.

46 In 2014, a laboratory prototype of a nanoscale logic machine, assembled using a variety of ingenious techniques, was reported by J. Ellenbogen and associates at Mitre and Harvard. http://www.pnas.org/content/111/7/2431.full.

47 Feynman, "Plenty of Room," 30, 34, 36.

48 Ed Regis, *Great Mambo Chicken and the Transhuman Condition* (Addison-Wesley, 1990), 119.

49 K. E. Drexler, *Engines of Creation: The Coming Era of Nanotechnology* (Anchor Press/ Doubleday, 1986).

50    Feynman, "Plenty of Room," 36.

51    Personal communication.

52    J. Storrs Hall, *Nanofuture* (Prometheus, 2005), p. 17.

53    As an aside, I will often (but not always) use the term nanotech to refer to the original, atomically precise, engineering vision, and nanotechnology to refer to the catchall nanoscale science and engineering, but mostly materials science, of the type funded under the auspices of the National Nanotechnology Initiative.

54    K. Eric Drexler, *Nanosystems: Molecular Machinery, Manufacturing, and Computation* (Wiley, 1992), 339.

55    For the basic capability to be expected, see *Nanosystems* Fig. 14.8: a one-kg system can produce one kg of product in an hour. Let it multiply exponentially for two days and you have 2.8e+14 machines, or 280 billion tonnes of nanomanufacturing equipment. The rule of thumb is that you replicate until you have 69 percent of the mass of the target product in replicators, and then set them all to work making the target. A more detailed estimate involves deciding on appropriate assumptions involving feedstocks, transportation, and so forth, but this will give you an idea of the raw creative power of an autogenous technology. *It is a possibility.*

56    From a Moderna press release, February 24, 2020.

57    Steven B. Krivit, *The Rebirth of Cold Fusion* (Pacific Oaks, 2004), 70. See also Eugene F. Mallove, *Fire From Ice* (Wiley, 1991), 40.

58    For a very rough (and extremely conservative) estimate, we can guess that there was as much heat produced as a 14-ounce Bernzomatic propane cylinder can provide: 18 megajoules. Roughly as much more energy would have been produced as neutrons and gammas as heat. Five sieverts, a lethal dose, comes to 500 J of absorbed radiation per 100-kg body. There would have been enough radiation, if evenly distributed, to kill 36,000 people.

59    Derived from Fig. 9a in Martin Fleischmann et al., "Calorimetry of the palladium-deuterium-heavy water system," *Journal of Electroanalytical Chemistry* 287 (1990): 293, https://citeseerx.ist.psu.edu/viewdoc/download?doi=10.1.1.380.1367&rep=rep1&type=pdf.

60    J. Rafelski and S.E. Jones, "Cold Nuclear Fusion," *Scientific American* 257 (July 1987): 84–89.

61    Krivit, *The Rebirth of Cold Fusion*, 72.

62    Fleischmann and Pons had wanted to end the title with a question mark, but bowed to pressure to omit it.

63    Richard Feynman, *The Meaning of It All* (Helix, 1998), 34.

64    Energy Research Advisory Board, *A Report of the Energy Research Advisory Board to the United States Department of Energy*, Washington, D.C., November 1989, DOE/S-0073 DE90 005611. A copy can be found at https://upload.wikimedia.org/wikipedia/commons/c/c6/Cold_Fusion_Research_-_ERAB_-_1989.pdf.

65    Charles G. Beaudette, *Excess Heat*, 2nd ed. (Oak Grove Press, 2002), 98.

66    Gustave C. Fralick, Arthur J. Decker, and James W. Blue, "Results of an Attempt to Measure Increased Rates of the Reaction D-2 + D-2 Yields He-3 + n in a Nonelectrochemical Cold Fusion Experiment," NASA TM 102430, 1989, https://ntrs.nasa.gov/search?highlight=true&q=TM-102430. Fralick had been looking for neutrons, which were not found, and so he didn't publish in the open literature, but merely filed the experiment's results as a NASA technical memo. The team did find excess heat, though, a temperature rise of 15° Celsius in the deuterium experiment, as compared to the light-hydrogen control.

67    Niccolò Machiavelli, *The Prince*, trans. William K. Marriott (Dutton, 1958), Chapter VI.

68    M. Miles and P. Hagelstein, "New Analysis of MIT Calorimetric Errors," *Journal of Condensed Matter Nuclear Science* 8 (2012): 132–138, https://iscmns.org/CMNS/JCMNS-Vol8.pdf.

69    This speech was delivered at the Atom Unexplored conference in Turin, Italy.

70    "Unreliable Research: Trouble at the Lab," *The Economist*, October 19, 2013, http://www.economist.com/news/briefing/21588057-scientists-think-science-self-correcting-alarming-degree-it-not-trouble—and don't even get me started on the psychology and other soft "sciences" literature.

71    That is to say, no positive results. Some of the Institute's experiments can be seen as establishing conditions and procedures that do not produce the effect.

72    "Personal Attorney of Pons, Fleischmann, Is Being Paid by the U," *Deseret News,* May 18, 1990, http://www.deseretnews.com/article/103053/PERSONAL-ATTORNEY-OF-PONS-FLEISCHMANN-IS-BEING-PAID-BY-THE-U.html?pg=all.

73    T. Roulette, J. Roulette, and S. Pons, "Results of ICARUS 9 Experiments Run at IMRA Europe," in *Sixth International Conference on Cold Fusion, Progress in New Hydrogen Energy,* (1996), https://www.lenr-canr.org/acrobat/RouletteTresultsofi.pdf.

74    Thomas W. Grimshaw, *Evidence-Based Public Policy Toward Cold Fusion: Rational Choices for a Potential Alternative Energy Source* (University of Texas at Austin, 2008).

75    Michael McKubre, personal communication.

76    A compilation of the reviewers' notes can be found at "2004 U.S. Department of Energy Cold Fusion Review Reviewer Comments," LNR-CANR, http://lenr-canr.org/acrobat/DOEusdepartme.pdf.

77    Peter W. Gorham, "Ballooning Spiders: The Case for Electrostatic Flight," November 2013, http://arxiv.org/pdf/1309.4731.pdf.

78    National Science and Technology Council, Committee on Technology, Interagency Working Group on Nanoscience, Engineering and Technology, "National Nanotechnology Initiative: Leading to the Next Industrial Revolution" (Washington, D.C., February 2000), 11, https://clintonwhitehouse4.archives.gov/media/pdf/nni.pdf.

79    Over 90 percent of the NNI budget is actually spent by the National Science Foundation, Department of Defense, Department of Education, and National Institutes of Health. Neal Lane and Thomas Kalil, "In the Beginning: The U.S. National Nanotechnology Initiative," in *Nanoethics: The Ethical and Social Implications of Nanotechnology,* ed. Fritz Allhoff, Patrick Lin, James Moor, and John Weckert (Wiley, 2007), 83.

80    In a lecture to the Newark College of Engineering, November 8, 1974.

81    For which, among other things, your humble narrator did a study of the applications of nanotechnology to flying cars.

82    "Nanotechnology: Drexler and Smalley Make the Case for and Against 'Molecular Assemblers,'" *Chemical and Engineering News,* December 1, 2003, http://pubsapp.acs.org/cen/coverstory/8148/8148counterpoint.html.

83    See, for example, K. Eric Drexler, David Forrest, Robert A. Freitas Jr., J. Storrs Hall, Neil Jacobstein, Tom McKendree, Ralph Merkle, Christine Peterson, "On Physics, Fundamentals, and Nanorobots: A Rebuttal to Smalley's Assertion That Self-Replicating Mechanical Nanorobots Are Simply Not Possible," Institute for Molecular Manufacturing, 2001, http://www.imm.org/publications/sciamdebate2/smalley. See also Drexler, *Radical Abundance* (PublicAffairs, 2013), Chapter 13.

84    J. Duarte et al., "Political Diversity Will Improve Social Psychological Science," *Behavioral and Brain Sciences* 38 (2015): E130, doi:10.1017/S0140525X14000430.

85    Then known as the United States National Museum.

86    "The Wright/Smithsonian Controversy: Patents and Politics," Wright Brothers Aeroplane Company, https://www.wright-brothers.org/History_Wing/History_of_the_Airplane/Doers_and_Dreamers/Wright_Smithsonian_Controversy/02_Patents_and_Politics.htm. Curtiss's suit was denied, and he ultimately bought out the Wrights' patent for $1,000,000, founding the Curtiss-Wright aircraft company.

87    "Wright Brothers," Wikipedia, http://en.wikipedia.org/wiki/Wright_brothers. Langley also got about $20,000 in private funding. In current dollars, that comes out to roughly $28,000 (Wrights) vs. $1,400,000 + $560,000 (Langley).

88    Terence Kealey, *The Economic Laws of Scientific Research* (Palgrave Macmillan, 1997).

89    OECD, *The Sources of Economic Growth in OECD Countries,* (OECD, 2003), 85.

90    The full title is *A Nation at Risk: The Imperative for Educational Reform,* the 1983 report of Ronald Reagan's National Commission on Excellence in Education.

91    Barry R. Chiswick, Nicholas Larsen, and Paul Pieper, "The Production of PhDs in the United States and Canada," IZA Discussion Paper Series no. 5367 (December 2010): 3–5.

92   Mark K. Fiegener, "Number of Doctorates Awarded in the United States Declined in 2010," National Center for Science and Engineering Statistics InfoBrief NSF 12-303 (November 2011): 2-3, http://www.nsf.gov/statistics/infbrief/nsf12303/nsf12303.pdf.

93   At the high end, at least. Increasing high-school dropout rates seem to match declining productivity figures somewhat better.

94   Kevin Drum, "The Great Stagnation," *Mother Jones*, January 31, 2011, https://www.motherjones.com/kevin-drum/2011/01/great-stagnation.

95   Daniel Dennett, *Consciousness Explained* (Boston: Back Bay Books, 1991), 177.

96   Defense Advanced Research Projects Agency, "Robots Conquer DARPA Grand Challenge," news release, October 8, 2005, http://archive.darpa.mil/grandchallenge05/GC05winnerv2.pdf.

97   "Topics of the Times," *New York Times*, January 13, 1920, 12, http://graphics8.nytimes.com/packages/pdf/arts/1920editorial-full.pdf.

98   "A Correction," *New York Times*, July 17, 1969, 43.

99   Dwight Eisenhower, "Farewell Address," January 17, 1961, transcript available at https://www.pbs.org/wgbh/americanexperience/features/eisenhower-farewell.

100  J. Storrs Hall, *Nanofuture* (Prometheus, 2005), 40.

101  H. G. Wells, *The Time Machine*, 1895, Chapter VI, https://www.gutenberg.org/files/35/35-h/35-h.htm.

102  Peter Turchin, *Ultrasociety: How 10,000 Years of War Made Humans the Greatest Cooperators on Earth* (Chaplin, CT: Beresta Books, 2015), 42.

103  George Orwell, "In Front of Your Nose," Tribune, March 22, 1946. The essay is also included in *George Orwell: Narrative Essays* (Harvill Secker, 2009). See https://www.orwellfoundation.com/the-orwell-foundation/orwell/essays-and-other-works/in-front-of-your-nose.

104  See, e.g., Robert Trivers, *Deceit and Self-Deception: Fooling Yourself the Better to Fool Others* (Allen Lane, 2011); and Kevin Simler and Robin Hanson, *The Elephant in the Brain: Hidden Motives in Everyday Life* (Oxford University Press, 2017).

105  Alison Flood, "TS Eliot Rejected Bloomsbury Group's 'Cursed Fund' to Work in Bank," *The Guardian*, September 2, 2009, https://www.theguardian.com/books/2009/sep/02/ts-eliot-bloomsbury-fund.

106  Simler and Hanson, *The Elephant in the Brain*, Chapters 13 and 14 in particular. See also Bryan Caplan, *The Case Against Education: Why the Education System Is a Waste of Time and Money* (Princeton University Press, 2018).

107  Rather than attempt to list these to show you how this effect works, I'll point you to an entire book full of them: Patrick Moore, *Fake Invisible Catastrophes and Threats of Doom* (2021). Note that Moore was a founder of Greenpeace.

108  The quoted passage is in John Burroughs, *Time and Change* (Riverside, 1912), 246.

109  Bill McKibben, *Eaarth: Making a Life on a Tough New Planet* (Henry Holt, 2010), xii.

110  Bill McKibben, *The End of Nature* (Anchor, 1989).

111  David M. Graber, "Mother Nature as a Hothouse Flower," *Los Angeles Times*, October 22, 1989, http://articles.latimes.com/1989-10-22/books/bk-726_1_bill-mckibben/2.

112  A. Leiserowitz et al., *Climate Change in the American Mind: May 2017*, Yale Program on Climate Change Communication (Yale University and George Mason University, 2017).

113  D. J. Arendt et al., "Key Economic Sectors and Services," in *Climate Change 2014: Impacts, Adaptation, and Vulnerability. Part A: Global and Sectoral Aspects*, AR 5, Intergovernmental Panel on Climate Change (2014), 662, https://www.ipcc.ch/site/assets/uploads/2018/02/WGIIAR5-Chap10_FINAL.pdf.

114  U.S. Government Accountability Office [formerly known as the General Accounting Office], "Climate Change: Information on Potential Economic Effects Could Help Guide Federal Efforts to Reduce Fiscal Exposure," Report GAO-17-720 (September 2017), https://www.energy.senate.gov/public/index.cfm?a=files.serve&File_id=90ECDE3F-38AF-47DB-8326-D38046D1E87E.

115  That is not to say that the science itself isn't biased and warped by the political priorities of the funding agencies; it is. But that is actually the most reassuring part of the busi-

ness: Even with literally billions of dollars of incentive to find the most alarming results possible, when you get down to the bottom line, the best they have managed to come up with is a 3 percent hit at the end of a century. There is no "crisis" here.

116   Michelle Nijhuis, "What Roads Have Wrought," *The New Yorker*, March 20, 2015, http://www.newyorker.com/tech/elements/roads-habitat-fragmentation.

117   . . . . such as Japan's TechnoFarm, which produces more lettuce per year in a one-acre building than one can harvest on a 300-acre outdoor farm: http://spread.co.jp/en/technofarm.

118   See, for example, J. B. Peires, *The Dead Will Arise: Nongqawuse and the Great Xhosa Cattle-Killing Movement of 1856-7* (Indiana University Press, 1989).

119   Carl Sagan, *The Demon-Haunted World: Science as a Candle in the Dark* (Ballantine Books, 1997), 25.

120   The Future History chart was first printed in *Astounding* in May 1941; it is reproduced on pp. iii–ix of *The Future History of Robert Heinlein, Vol I*, which is Volume XXII of the Virginia Edition of Heinlein's collected works.

121   Wendell Cox and Jean Love, "40 Years of the US Interstate Highway System: An Analysis; The Best Investment a Nation Ever Made," American Highway Users Alliance, June 1996, http://www.publicpurpose.com/freeway1.htm.

122   Throughout the book, we will use "convertible" to mean an airplane that converts to a car (and not a car with a folding roof). The term "roadable" is also used, but its implications don't fit exactly with the kind of vehicle we are driving at.

123   Second Continental Congress, *The Unanimous Declaration of the Thirteen United States of America*, 1776.

124   Austin B. Hallock, *Sky Full of Dreams: The Aviation Exploits, Creations, and Visions of Bruce K. Hallock* (Elevon Books, 2010), Chapters 7 and 8.

125   Andrew Glass, *Flying Cars, the True Story* (Clarion Books, 2015), Chapter 13.

126   "The Ups and Downs of the Flying Car," *Reason*, August 1984, 40–42, http://www.unz.org/Pub/Reason-1984aug-00040.

127   National Highway Traffic Safety Administration, *PB 211–015: Evaluation of the 1960–1963 Corvair Handling and Stability* (July 1972).

128   One such scholar is Glenn Reynolds, in "Why We Still Don't Have Flying Cars," *USA Today*, May 12, 2016, http://www.usatoday.com/story/opinion/2016/05/12/technological-progress-stagnation-regulatory-explosion-1970s-column/84225066.

129   W. David Montgomery et al., "Macroeconomic Impacts of Federal Regulation of the Manufacturing Sector," NERA Economic Consulting, commissioned by MAPI, August 2012, https://www.ohiomfg.com/grip-assets/resource_library/environment_management/nera_mapi_regulationsreport_august2012.pdf.

130   Federal Air Regulations Part 91, section 1443, paragraph (c).

131   John W. Dawson and John J. Seater, "Federal Regulation and Aggregate Economic Growth," *Journal of Economic Growth* 18 (2013): 137–177.

132   S. Peltzman, "An Evaluation of Consumer Protection Legislation: 1962 Drug Amendments," *Journal of Political Economy* 81, no. 5 (September–October 1973): 1049–1091, https://www.journals.uchicago.edu/doi/abs/10.1086/260107.

133   Alfred E. Kahn, "Airline Deregulation," *Concise Encyclopedia of Economics*, Library of Economics and Liberty, n.d., http://www.econlib.org/library/Enc1/AirlineDeregulation.html.

134   The Cessna model 172 Skyhawk is the most widely produced light aircraft in history.

135   "The Cessna Story," 172Guide, n.d., https://www.172guide.com/history.htm.

136   W. Kip Viscusi, "Liability," *Concise Encyclopedia of Economics*, Library of Economics and Liberty, n.d., http://www.econlib.org/library/Enc/Liability.html.

137   Ibid.

138   The cost of malpractice insurance varies wildly with specialty and geography, ranging roughly from $20,000 to $200,000 a year. It's actually not as bad now as it was last century, since many states (33 at last count) have passed liability-limitation laws.

139 "Tillinghast Study: U.S. Tort Costs Reach a Record $260 Billion," *Insurance Journal*, March 13, 2006, https://www.insurancejournal.com/news/national/2006/03/13/66411.htm.

140 David R. Henderson, "German Economic Miracle," *Concise Encyclopedia of Economics*, Library of Economics and Liberty, n.d., https://www.econlib.org/library/Enc/GermanEconomicMiracle.html.

141 Robert A. Heinlein, "Notebooks of Lazarus Long," *Analog Science Fiction/Science Fact*, June 1973, 77.

142 Henderson, "German Economic Miracle."

143 H. G. Wells, *Anticipation* (Chapman & Hall, 1902), https://www.gutenberg.org/ebooks/19229.

144 D. S. L. Cardwell, *Turning Points in Western Technology* (SHP/Neale Watson, 1972), 198. Emphasis in the original.

145 By 2000, modern plants produced 75 cars per worker per year, another almost-but-not-quite factor of four.

146 See his book *Factfulness: Ten Reasons We're Wrong About the World—and Why Things Are Better Than You Think* (Flatiron Books, 2018).

147 In the real world, virtually all pilots use GPS, iPads, and other modern electronic devices. But you are required to learn to do it the old-fashioned way, just in case.

148 https://worldwarwings.com/10-wwii-aircraft-facts/ See also Marlyn R. Pierce, "Earning Their Wings: Accidents and Fatalities in the United States Army Air Forces During Flight Training in World War", Ph.D. dissertation, Kansas State University, 2013.

149 Well, almost. The pressure on the top is actually 1.28 ounces less than that on the bottom. But, by a strange coincidence, the air inside the box just happens to weigh 1.28 ounces, canceling the unbalanced force. If you replaced that air with helium, weighing only 0.18 ounces, you would get 1.1 ounces of lift.

150 If you ever hear anyone arguing about Newton vs. Bernoulli as a proper explanation of lift, please remember that, as valid physics, both must be true at the same time. The pressure patterns of accelerating flows cause the momentum change in the airstream, or vice versa. The two descriptions are inseparable.

151 Simulated with the SU2 open-source fluid dynamics program at Mach 0.2, and visualized with the open-source ParaView viewer.

152 Ira H. Abbott and Albert E. von Doenhoff, *Theory of Wing Sections* (Dover Publications, 1959), 538–9.

153 In an airliner you are cruising at an altitude around 36,000 feet, but the plane is pressurized to the equivalent of about 9,000. Anoxia is not a problem, but dehydration still is.

154 Paul Hoversten, "Why Do Helicopter Pilots Sit in the Right Seat?," *Smithsonian Air & Space Magazine*, November 16, 2011, https://www.airspacemag.com/need-to-know/why-do-helicopter-pilots-sit-in-the-right-seat-243212.

155 In 1931, Jim Ray (Pitcairn's test pilot) took off from the White House lawn in a PCA-2, needing only a measured 44 feet. On an unobstructed rooftop pad, you would only need enough speed to hold altitude at the far edge, rather than enough to do a steep climb-out.

156 Andreas Schäfer, "Regularities in Travel Demand: An International Perspective," *Journal of Transportation and Statistics* 3, no. 3 (December 2000), https://citeseerx.ist.psu.edu/viewdoc/download?doi=10.1.1.454.5721&rep=rep1&type=pdf. Graph courtesy Andreas Schäfer.

157 The technical specs can be found at http://www.boeing.com/commercial/747family/specs.html and https://modernairliners.com/boeing-747-jumbo/boeing-747-specs.

158 Since I originally researched this chapter, oil and gas prices have dropped significantly: Jet A typically costs from $4–$5 per gallon at the moment, ranging from $3.25 in Ocean City, MD, to $7.69 at Reagan National Airport. Even so, it still costs hundreds of thousands of dollars to fuel the 747.

159 Thousands of cubic feet. One MCF of natural gas produces about a million BTU of heat.

160  "Frequently Asked Questions (FAQs): What Do I Pay for in a Gallon of Gasoline and Diesel Fuel?," U.S. Energy Information Administration, Independent Statistics & Analysis, September 11, 2020, http://www.eia.gov/tools/faqs/faq.cfm?id=22&t=6.

161  No, of course not exactly. You would have to use each fuel in perfectly even proportions for a simple average to hold. But the bottom line is true: Chemical fuels are, in general, about 1,000 times as expensive as nuclear for the same energy.

162  A megawatt diesel generator burns about 80 gallons, or 500 pounds, of fuel per hour. A megawatt wind turbine, by amusing coincidence, goes through about the same amount of lubricating oil per year. But a megawatt engine burning uranium would consume less than one pound of it in a year.

163  A one-foot-cube battery could run the lights on your car, but not the car itself, according to "Prototype Battery Packs 10 Times More Power," Moscow Institute of Physics and Technology, May 31, 2018, https://mipt.ru/english/news/prototype_nuclear_battery_packs_10_times_more_power.

164  Mark Halper, "The U.S. Is Helping China Build a Novel, Superior Nuclear Reactor," *Fortune*, February 2, 2015, https://fortune.com/2015/02/02/doe-china-molten-salt-nuclear-reactor.

165  "2011 Tōhoku Earthquake and Tsunami: Damage and Effects," Wikipedia, accessed 2016, https://en.wikipedia.org/wiki/2011_Tohoku_earthquake_and_tsunami#Damage_and_effects. The heroic older workers who were allowed to enter the plant for emergency repairs, and whose dose limits were higher than normal, are expected to have a 1 percent greater risk of cancer.

166  United Nations Scientific Committee on the Effects of Atomic Radiation, *Sources, Effects and Risks of Ionizing Radiation: 2020 Report*, Scientific Annex B (United Nations, 2020), 107, http://www.unscear.org/unscear/en/fukushima.html.

167  James Conca, "How Deadly Is Your Kilowatt? We Rank the Killer Energy Sources," *Forbes*, June 10, 2012, http://www.forbes.com/sites/jamesconca/2012/06/10/energys-deathprint-a-price-always-paid.

168  Herman Kahn and Anthony J. Wiener, *The Year 2000: A Framework for Speculation on the Next Thirty-Three Years*, (Macmillan, 1967), 71.

169  Olivier Deschenes and Enrico Moretti, "Extreme Weather Events, Mortality and Migration," NBER Working Paper No. 13227, July 2007, http://www.nber.org/papers/w13227.

170  Maurice Tubiana et al., "The Linear No-Threshold Relationship Is Inconsistent With Radiation Biologic and Experimental Data," *Radiology* 251, no. 1 (April 2009): 13–22, http://www.ncbi.nlm.nih.gov/pmc/articles/PMC2663584.

171  "Backgrounder on Biological Effects of Radiation," United States Nuclear Regulatory Committee, March 2017, http://www.nrc.gov/reading-rm/doc-collections/fact-sheets/bio-effects-radiation.html.

172  And yes, it was almost entirely regulation: See J. Lovering, A. Yip, and T. Nordhaus, "Historical Construction Costs of Global Nuclear Power Reactors," *Energy Policy* 91 (April 2016): 371–382, https://doi.org/10.1016/j.enpol.2016.01.011.

173  "Nuclear Power Learning and Deployment Rates; Disruption and Global Benefits Forgone," *Energies* 10, no. 12 (2017): 2169, doi:10.3390/en10122169.

174  Charles Barton, "Sherrell Greene on Liquid Chloride Reactors, 'Business as Usual,' and a Second Manhattan Project," *The Nuclear Green Revolution*, November 19, 2011, http://nucleargreen.blogspot.com/2011/11/sherrell-greene-on-liquid-chloride.html.

175  John Lehman, "Obama Torpedoes the Nuclear Navy," *Wall Street Journal*, May 25, 2015, http://www.wsj.com/articles/obama-torpedoes-the-nuclear-navy-1432591747.

176  Ansgar Belke, Christian Dreger, and Frauke de Haan, "Energy Consumption and Economic Growth: New Insights into the Cointegration Relationship," Deutsches Institut für Wirtschaftsforschung Discussion Paper 1017, June 2010, http://www.diw.de/documents/publikationen/73/diw_01.c.357400.de/dp1017.pdf. "Granger causality" is a statistical test showing that one data set is useful in predicting another.

177   David C. Rode, et al., "The Retirement Cliff: Power Plant Lives and Their Policy Implications," *Energy Policy* 106 (July 2017): 222–232, https://doi.org/10.1016/j.enpol.2017.03.058.

178   Arthur C. Clarke, *Profiles of the Future* (Chaucer Press revised edition, 1973), 170.

179   Ibid, 161.

180   Kenneth S. Krane, *Introductory Nuclear Physics* (Wiley, 1988), 2. This is a rewrite and update of the even more classic David Halliday text of the same title from 1955.

181   Wade Marcum and Bernard Spinrad, "Nuclear Reactor," *Encyclopedia Britannica*, September 5, 2019, http://www.britannica.com/EBchecked/topic/421763/nuclear-reactor/302452/History-of-reactor-development.

182   All three remarks are widely quoted, but appear together in Paul Ciotti, "Fear of Fusion: What If It Works?," *Los Angeles Times*, April 19, 1989.

183   Mike Gray, *Angle of Attack: Harrison Storms and the Race to the Moon* (Penguin, 1994) 145–149.

184   Paul Alivisatos et al., *Productive Nanosystems: A Technology Roadmap*, (Battelle Memorial Institute and Foresight Nanotech Institute, 2007), https://www.foresight.org/roadmaps/Nanotech_Roadmap_2007_main.pdf.

185   George Friedman, "The Space Studies Institute View on Self-Replication," *Newsletter of the Molecular Manufacturing Shortcut Group of the National Space Society* 4, no. 4 (1996), http://www.islandone.org/MMSG/9612.html.

186   Feynman, "Plenty of Room," 34.

187   Robert Freitas and Ralph Merkle, *Kinematic Self-Replicating Machines* (Landes Bioscience, 2004), 116. Full text available online: http://www.molecularassembler.com/KSRM.htm.

188   Lars Pilø, "The Reconstruction of the Lendbreen Tunic," Secrets of the Ice, June 28, 2016, http://secretsoftheice.com/funn/2016/06/28/reconstruction.

189   Adam Smith, *An Inquiry Into the Nature and Causes of the Wealth of Nations* (1776), Chapter I, https://www.gutenberg.org/files/3300/3300-h/3300-h.htm.

190   "Space and the City," *The Economist*, April 4, 2015, https://www.economist.com/leaders/2015/04/04/space-and-the-city.

191   Christopher Morrison, "Extrasolar Object Interceptor and Sample Return Enabled by Compact, Ultra Power Dense Radioisotope Batteries," NASA, Feb 25, 2021, https://www.nasa.gov/directorates/spacetech/niac/2021_Phase_I/Extrasolar_Object_Interceptor_and_Sample_Return.

192   J. Storrs Hall, *Beyond AI: Creating the Conscience of the Machine* (Prometheus, 2007), 260.

193   Will Knight, "This Factory Robot Learns a New Job Overnight," *MIT Technology Review*, March 18, 2016, https://www.technologyreview.com/s/601045/this-factory-robot-learns-a-new-job-overnight.

194   See Andrej Karpathy, "The Unreasonable Effectiveness of Recurrent Neural Networks," Andrej Karpathy Blog, May 21, 2015, http://karpathy.github.io/2015/05/21/rnn-effectiveness.

195   Hall, *Beyond AI*, 170.

196   Cf. David Silver et al., "Mastering the Game of Go With Deep Neural Networks and Tree Search," *Nature* 529, (January 28, 2016): 484–489, doi:10.1038/nature16961.

197   Sam Adams, Itmar Arel, Joscha Bach, Robert Coop, Rod Furlan, Ben Goertzel, J. Storrs Hall, Alexei Samsonovich, Matthias Scheutz, Matthew Schlesinger, Stuart C. Shapiro, John Sowa, "Mapping the Landscape of Human-Level Artificial General Intelligence," *AI Magazine* 33, no. 1 (2012), http://dx.doi.org/10.1609/aimag.v33i1.2322. My contribution was, you guessed it, the Wozniak Test.

198   N. J. Nilsson, "Human-Level Artificial Intelligence? Be Serious!," *AI Magazine* 26, no.4 (2005): 68, https://doi.org/10.1609/aimag.v26i4.1850.

199   You can cut these numbers roughly in half if you use either a statistical model or a game-playing engine for probabilistic compression, but that would imply you already knew something about strategy. Here we are concerned with how simple the rules make the game for a program attempting to learn strategy from scratch. Note that DeepMind followed up AlphaGo with AlphaZero, a program that did just that.

200  Jack Clark, "Why Google Wants to Sell Its Robots: Reality Is Hard," *Bloomberg*, March 18, 2016, http://www.bloomberg.com/news/articles/2016-03-18/why-google-wants-to-sell-its-robots-reality-is-hard.

201  U.S. Bureau of Labor and Statistics data; thanks to Mark Perry for the pointer.

202  It's 45373 $e^{0.0114t+0.0001116t^2}$, where $t$ is year–1950.

203  J.M. Keynes, "Economic possibilities for our grandchildren," Essays in Persuasion (Norton, 1963), 358–73. The piece was originally published in 1930.

204  See the Federal Reserve Bank of St. Louis's data, at https://fred.stlouisfed.org/, for the series MANEMP and CE16OV.

205  "To you, a robot is a robot. Gears and metal; electricity and positrons. Mind and iron! Human-made! If necessary, human-destroyed! But you haven't worked with them, so you don't know them. They're a cleaner, better breed than we are." Isaac Asimov, *I, Robot*, (Gnome, 1950), Introduction.

206  Elaine Pofeldt, "This Crime in the Workplace Is Costing US Businesses $50 Billion a Year," CNBC, September 12, 2017, https://www.cnbc.com/2017/09/12/workplace-crime-costs-us-businesses-50-billion-a-year.html.

207  Of course, there are plenty of psychopaths among the human race, and all of them want to get their hands on any powerful technology they can, including AI. But that is all the more reason for the 98 percent or so of us who are not psychopaths to build sanely de-signed AI to detect, contain, and counteract them.

208  And when I say "we," I mean robot computer scientists, all of whom will have read all my books.

209  Stable enough with a 900-year half-life, it has a thermal neutron capture cross section of 5,000 barns (vs., e.g., 585 for U-235). Note that this is only the moderated reaction: *Popular Science* in 1961 told us that californium could be used to make bullet-size atom bombs, but the unmoderated (fast-neutron) critical mass of Cf is more like 11 pounds.

210  No, you couldn't extract the plutonium to make a bomb; the smallest steel-reflected non-moderated (i.e., the bomb reaction) critical mass of Pu-241 is over 11 pounds. See "Critical Mass," European Nuclear Society, n.d., https://www.euronuclear.org/glossary/critical-mass.

211  "Small Nuclear Power Reactors," World Nuclear Association, June 2021, https://www.world-nuclear.org/information-library/nuclear-fuel-cycle/nuclear-power-reactors/small-nuclear-power-reactors.aspx.

212  J. Stephen Herring, Stephen Mackwell, Christopher Pestak, and Karin Hilser, "Small Modular Fission Reactors for Space Applications," *Nuclear and Emerging Technologies for Space, American Nuclear Society Topical Meeting* (2019), 6, http://anstd.ans.org/NETS-2019-Papers/Track-4--Space-Reactors/abstract-119-0.pdf.

213  "Lattice Confinement Fusion," NASA, n.d., https://www1.grc.nasa.gov/space/science/lattice-confinement-fusion.

214  Polonium-210's half-life of 138 days is something of a "sweet spot," making it last long enough to use but giving it a high enough level of radioactivity to be very toxic. Po-210 emits high-energy (5.3 MeV) alphas; for comparison, the betas emitted by tritium are 0.018 MeV.

215  Michael Koziol, "Whether Cold Fusion or Low-Energy Nuclear Reactions, U.S. Navy Researchers Reopen Case." *IEEE Spectrum*, March 22, 2021, https://spectrum.ieee.org/tech-talk/energy/nuclear/cold-fusion-or-low-energy-nuclear-reactions-us-navy-researchers-reopen-case.

216  What he actually said, in a 1926 letter to Max Born, was, "The theory produces a good deal but hardly brings us closer to the secret of the Old One. I am at all events convinced that He does not play dice."

217  As quoted by Heisenberg in *Physics and Beyond* (Harper & Row, 1971), 206.

218  Richard Feynman, *The Character of Physical Law* (MIT, 1995), 129. Also, you can watch him say it at https://www.youtube.com/watch?v=w3ZRLllWgHI.

219  Or, if you want to get deeper into the structure, this pushes the positively charged Up quarks one way, and pulls the negatively charged Down quarks the other. But nobody has

ever done a simulation of even a milligram of loaded palladium in quantum electrodynamics without using the Born-Oppenheimer approximation, in which the nuclei are
treated classically, much less a quantum chromodynamic one.

220  The patents are No. 5,928,483, "Electrochemical Cell Having a Beryllium Compound
     Coated Electrode," and No. 8,419,919, "System and Method for Generating Particles."

221  H. Hora, S. Eliezer, G. Kirchhoff, N. Nissim, J. Wang, et al., "Road Map to Clean Energy
     Using Laser Beam Ignition of Boron-Hydrogen Fusion," *Laser and Particle Beams* 35, no.
     4 (December 2017): 730–740, doi:10.1017/S0263034617000799.

222  You can see the fan's specs at https://www.schuebeler-jets.de/pdf/Datenblaetter/
     EN/215HST_Flyer-ENG.pdf. Note that the compact form costs you in efficiency; a mod-
     ern electric motor driving a 32-inch prop can get 15 pounds of thrust per kW instead
     of 3.6.

223  This would be less than a third of the actual energy in the avgas, but the electric motors
     use it more efficiently.

224  The companies were Eviation and Lilium—see *AOPA Pilot*, June 2021, p. 31.

225  The fire occurred on April 17, 2021, in Spring, Texas: Tom Abrahams and Jeff Ehling, "Elon
     Musk Claims Autopilot Was Not Used in Fiery Tesla Crash That Killed 2 People in the
     Woodlands," ABC, April 19, 2021, https://abc13.com/2-killed-in-fiery-tesla-crash-that-
     took-4-hours-to-extinguish/10525148.

226  Jennifer Chu, "A Mighty Wind," *MIT News*, April 3, 2013, http://web.mit.edu/newsof-
     fice/2013/ionic-thrusters-0403.html.

227  Probably. See https://quoteinvestigator.com/2015/03/24/mousetrap.

228  Arthur C. Clarke, *Childhood's End*, (Pan, 1956), 62.

229  The incident occurred on September 26, 2014, and is detailed in "Chicago Center Fire
     Contingency Planning and Security Review," Federal Aviation Authority, n.d., https://
     www.faa.gov/news/media/ZAU_Fire_Public_Review.pdf.

230  FAA/NASA Unmanned Aircraft System Traffic Management (UTM) Research Transition
     Team Concept Working Group, Concept & Use Cases Package #2: Technical Capability
     Level 3, Version 1.0, March 2018, 8, https://ntrs.nasa.gov/api/citations/
     20190004908/downloads/20190004908.pdf.

231  Frederick I. Ordway III, "*2001: A Space Odyssey* in Retrospect," The Kubrick Site, n.d.,
     http://www.visual-memory.co.uk/amk/doc/0075.html.

232  There is a newly discovered low-energy trajectory, called the ballistic capture transfer,
     which takes about as long but avoids the launch window restriction. Francesco Topputo
     and Edward Belbruno, "Earth–Mars Transfers with Ballistic Capture," *Celestial Mechanics
     and Dynamical Astronomy* 121 (2015): 329–346, https://link.springer.com/article/10.1007/
     s10569-015-9605-8.

233  A gigameter is a million kilometers—150 Gm is the same as one astronomical unit, or AU,
     the average distance between Earth and the Sun.

234  The Davy Crockett hand-launched nuclear bazooka is generally thought of as the small-
     est, but Taylor claimed later that an entire implosion weapon that would fit into a six-inch
     ball had been built. The S.O.B., or Super Oralloy Bomb, was our biggest non-hydrogen
     bomb; it had a yield of half a megaton. (Oralloy is "Oak Ridge alloy," i.e., highly enriched
     U-235.)

235  Quoted in George Dyson, *Project Orion: The True Story of the Atomic Spaceship* (Henry
     Holt, 2002), 105.

236  E.g., the Celebrity Reflection.

237  Dyson, *Project Orion*, 230.

238  . . . assuming that the average person had 40 years to live, and using U.S. cancer inci-
     dence rates, as documented by the National Cancer Institute: https://www.cancer.gov/
     about-cancer/understanding/statistics.

239  Last I looked, COVID-19 had killed about four million people. If it had had the virulence of
     the bubonic plague, it would have killed some 2,500 million.

240  Carl Sagan and Ann Druyan, *Shadows of Forgotten Ancestors* (Ballantine Books,
     1993), 133.

241 Carl Sagan, *Pale Blue Dot: A Vision of the Human Future in Space* (Ballantine Books, 1997), 266.

242 See an introductory analysis at http://www.ce.jhu.edu/perspectives/studies/Eiffel percent-20Tower percent20Files/ET_Introduction.htm.

243 "Wings of Steel," *The Economist*, February 7, 2015, http://www.economist.com/news/science-and-technology/21642107-alloy-iron-and-aluminium-good-titanium-tenth. Note that in popular discussions of this and similar advances, the term "specific strength" is often misinterpreted as simply meaning strength. It means the strength to weight ratio.

244 Philip Francis Nowlan, *The Airlords of Han*; see also *Armageddon 2419 A.D.* First published in *Amazing Stories* March 1929 and August 1928, respectively. Both are available at Project Gutenberg: https://www.gutenberg.org/ebooks/author/31334.

245 Megan McArdle, "Real New Yorkers Can Say Goodbye to All That," *Bloomberg*, June 1, 2015, http://www.bloombergview.com/articles/2015-06-01/real-new-yorkers-can-say-goodbye-to-all-that.

246 Edward Elmer Smith, *First Lensman* (Fantasy Press, 1952), 247.

247 Remember how air travel had a rate of 0.05 deaths per billion kilometers traveled and driving had a rate of 3.1? The corresponding figure for walking is 54.2.

248 Lawrence Solomon, "I Knew Jane Jacobs, and Everything Urban Lefties Say About Her Is Wrong," May 13, 2016, *Financial Post*, https://financialpost.com/opinion/i-knew-jane-jacobs-and-everything-urban-lefties-say-about-her-is-wrong.

249 For the promises, see, for example, Robert Heinlein, *The Roads Must Roll* (1940); Arthur C. Clarke, *Against the Fall of Night* (1948); Isaac Asimov, *The Caves of Steel* (1953).

250 Sarah Knapton, "Move to a Leafy Suburb to Cut Cancer Risk," *Telegraph*, April 14, 2016, http://www.telegraph.co.uk/science/2016/04/14/move-to-a-leafy-suburb-to-cut-cancer-risk.

251 E.g., Yehoshua Kolodny and I.R. Kaplan, "Uranium Isotopes in Sea-Floor Phosphorites, *Geochimica et Cosmochimica Acta* 34, no. 1 (January 1970): 3–24.

252 Arthur C. Clarke, *Profiles of the Future* (Chaucer Press revised edition, 1973), 74.

253 "Cosmological Constant Problem," Wikipedia, accessed July 22, 2021, https://en.wikipedia.org/wiki/Cosmological_constant_problem.

254 Carver Mead, *Collective Electrodynamics* (MIT, 2000), 1.

255 Widely noted, e.g., in Mead's *American Spectator* interview (September/October 2001, Vol. 34 Issue 7, p. 68) and in Mara Beller, *Quantum Dialogue: The Making of a Revolution* (Chicago, 1999), 126.

256 John Cramer, *The Quantum Handshake* (Springer, 2016), and earlier papers. Cramer developed Wheeler and Feynman's "absorber theory" into a full-fledged interpretation of QM, often known as the transactional interpretation. He would have been as good a giant for shoulders as Mead, but I didn't know him.

257 J. A. Wheeler and R. P. Feynman, "Interaction With the Absorber as the Mechanism of Radiation," *Reviews of Modern Physics* 17, no. 2–3 (April 1945): 157–181; and J. A. Wheeler and R. P. Feynman, "Classical Electro-dynamics in Terms of Direct Interparticle Action," *Reviews of Modern Physics* 21, no. 3 (July 1949): 425–433. I remain flabbergasted at how understandable physics papers published before 1950 are, compared to ones written today.

258 J. Storrs Hall, "A Space Pier," paper presented to the International Space Development Conference (ISDC), Washington, D.C., May 16, 2005. Note that further work has been done on the concept since that presentation.

259 In fact, they have been tested at orbital speeds (in a vacuum chamber fed by a particle accelerator!) by the European Space Agency: "World-First Firing of Air-Breathing Electric Thruster," European Space Agency, May 3, 2018, http://www.esa.int/Our_Activities/Space_Engineering_Technology/World-first_firing_of_air-breathing_electric_thruster.

260 You should not be surprised to find that my views on this subject follow fairly closely the work of William Nordhaus, who won the Nobel Prize in Economics for a lifetime of work on it.

261  Actually, it ranges from 300 over the tropics to 200 over the Great White North.
262  In simplistic classical physics. In reality, trying to put that much energy into that little space would do things that would make the Large Hadron Collider look like a lightning bug.
263  This is from a letter to Joseph Priestley, a British chemist and philosopher. The full text can be found on the website of the National Archives, at https://founders.archives.gov/documents/Franklin/01-31-02-0325.
264  Jill Lepore, "A Golden Age for Dystopian Fiction," *The New Yorker*, May 29, 2017.
265  John Stuart Mill, *Autobiography* (Oxford, 1873), 290.
266  Thomas Babington Macaulay, "Review of Southey's Colloquies on Society," *Edinburgh Review*, January 1830.
267  Robert Heinlein (under the pseudonym Anson MacDonald), "Beyond This Horizon," published in two parts in *Astounding*, April and May 1942. Available as a single-volume edition from Fantasy Press, 1948. The quoted passage is the opening paragraph.
268  In a letter to his brother Charles.
269  Henry Adams, *The Education of Henry Adams* (1907), 496, https://www.gutenberg.org/files/2044/2044-h/2044-h.htm.

# Index

# Acknowledgments

I have listed in the text many of the great science fiction writers and futurists who formed the zeitgeist of the future in which I grew up. Personal influences and valuable help were also many. In my understanding of AI and computer science, Charles Lytle, Charles Hedrick, Marvin Minsky, Ray Solomonoff, Trenchard More, Lou Steinberg, Al Despain, Irv Rabinowitz, and Bob Webber were seminal friends and mentors. In the field of nanotechnology, Eric Drexler, Ralph Merkle, Rob Freitas, Damian Allis, David Forrest, Tom McKendree, Chris Peterson, Al Globus, and Mark Sims were indispensable. Charles Simanski taught me to fly a plane, and Bob Snyder taught me to pilot a gyro. I learned a lot about cold fusion from Michael McKubre and Peter Hagelstein, who also read an earlier version of this manuscript. Other readers and reviewers were Jeff Hamilton, Larry Hudson, Peter McCloskey, Robin Hanson, Ralph Merkle, Rob Freitas, Ben Reinhardt, Alex Epstein, Jason Crawford, T. J. Snodgrass, and Tyler Cowen, all of whom offered valuable advice and encouragement. Without the efforts of my editor Thomas LeBien, this version would be much less readable; and without the hard work and support of my wife, Sandy Hall, it couldn't have happened at all.

# About the Author

J. Storrs Hall, Ph.D., is an independent scientist and author. He was the founding chief scientist of Nanorex, Inc., and a president of the Foresight Institute, and is currently a research fellow at the Institute for Molecular Manufacturing and an associate editor of the *International Journal of Nanotechnology and Molecular Computation*. He was also accredited as an expert reviewer for the Intergovernmental Panel on Climate Change in the field of computational climate models. His previous books include *Beyond AI: Creating the Conscience of the Machine* and *Nanofuture: What's Next for Nanotechnology*. Now residing on Chesapeake Bay, he dabbles in aerodynamics design under the auspices of Eastern Shore Flying Cars, LLC.

**Notes from the Art Department**

Creative Director: Tyler Thompson
Typesetting & Design: Kevin Wong
Printing: Hemlock Printers Ltd.
Bindery: Roswell Bookbinding

Composed in Teodor, designed by
Martin Vácha and Damian Nika
Langoz; Roobert, designed by Martin
Vácha, both from Displaay; and Record
Gothic Mono by A2 Type Co.

Printed on 70# Lynx

Bound in Euro Buckram

Inks:
■ Black
■ Pantone 7460 U

Stripe
Press